VOLUME THIRTY TWO

THE ENZYMES

Eukaryotic RNases and their Partners in RNA Degradation and Biogenesis, Part B

VOLUME THIRTY TWO

THE ENZYMES

Eukaryotic RNases and their Partners in
RNA Degradation and Biogenesis, Part B

Edited by

FENG GUO

Department of Biological Chemistry,
David Geffen School of Medicine,
Molecular Biology Institute,
University of California,
Los Angeles, CA 90095, USA

FUYUHIKO TAMANOI

Department of Microbiology, Immunology,
and Molecular Genetics, Molecular Biology Institute
University of California
Los Angeles, CA 90095, USA

AMSTERDAM • BOSTON • HEIDELBERG • LONDON
NEW YORK • OXFORD • PARIS • SAN DIEGO
SAN FRANCISCO • SINGAPORE • SYDNEY • TOKYO
Academic Press is an imprint of Elsevier

Academic Press is an imprint of Elsevier
225 Wyman Street, Waltham, MA 02451, USA
525 B Street, Suite 1900, San Diego, CA 92101-4495, USA
Radarweg 29, PO Box 211, 1000 AE Amsterdam, The Netherlands
The Boulevard, Langford Lane, Kidlington, Oxford, OX51GB, UK
32, Jamestown Road, London NW1 7BY, UK

First edition 2012

ISBN: 978-0-12-404741-9
ISSN: 1874-6047

For information on all Academic Press publications
visit our website at store.elsevier.com

Printed and bound in USA

12   13   14   15   11   10   9   8   7   6   5   4   3   2   1

CONTENTS

PREFACE

Study of small RNAs (20–30 nucleotides in length) has made a big impact on biomedical research in the past 15–20 years. MicroRNAs (miRNAs) are endogenous non-protein-coding RNAs that are extensively involved in development and cell physiology. Small interfering RNAs (siRNAs) are central to RNA interference that serves as an RNA-based immune system. The RNAi technology has transformed how gene functions are studied and is being explored as therapeutics. siRNAs may also be generated from double-stranded RNAs transcribed from an organism's own genome, and these endogenous siRNAs are involved in transposon suppression and heterochromatin formation. PIWI-interacting RNAs (piRNAs) maintain genome stability in germ cells by silencing transposon elements.

While these small RNAs differ from each other in their sequences, biogenesis, and functions, they all require nucleases and RNA-binding proteins for maturation and for regulating the expression of target genes. In this volume of *The Enzymes*, we present most up-to-date reviews of the nucleases and nucleic acid-binding proteins that play critical roles in processing the small RNAs from their usually much-longer precursors and in mediating target gene silencing. Chapter 1 summarizes how the ribonuclease III enzyme Dicer cleaves both miRNA and siRNA precursors and facilitates the assembly of the RNA-induced silencing complexes (RICS). Chapter 2 describes how three *Drosophila* RNA-binding proteins, produced from a single gene *loquacious* via alternative splicing, partner specifically with either Dicer-1 or Dicer-2 to direct the cleavage of miRNAs and siRNAs. In Chapter 3, how a heteromeric complex C3PO, composed of TRAX and translin, activates the RISC activity is discussed in the context of the crystal structure of octomeric C3PO, which reveals a curious football-shaped cage with the RNA-binding residues and ribonuclease active sites located in the interior. Chapter 4 presents the recent structural studies of the central component of RISC, the Argonaute proteins.

Whereas the first four chapters concern proteins that mainly function in the cytoplasm, the following two chapters discuss the small RNA processing events in the nucleus. Chapter 5 focuses on how another ribonuclease III enzyme Drosha and its RNA-binding partner DGCR8 work together to recognize and cleave primary transcripts of miRNAs. Chapter 6 reviews how the SMAD proteins, critical components of the TGFβ signaling pathway, regulate the Drosha-mediated processing of a subset of miRNAs.

Chapter 7 explains how the PIWI proteins collaborate with other nucleases and use its own RNase H-like slicer activity to contribute to the production of piRNAs in the nucleus and cytoplasm. Chapter 8 describes how the tumor suppressor p53, often described as "the guardian of the genome," and the MCPIP1 ribonuclease control miRNA maturation at the Drosha and Dicer cleavage steps, respectively.

Chapter 9 addresses the most critical technical hurdle that is holding back the application of RNA interference technology for therapeutic purposes—how to deliver small RNAs to where they need to go in combating diseases. Nanoparticle-based delivery methods and their applications in cancer treatment are reviewed.

There are topics that we have not been able to cover in this volume. We hope to be able to discuss other critical topics related to small RNAs in future volumes. We thank all the authors for their commitment and contribution to this volume. We also thank Mary Ann Zimmerman for her encouragement and guidance with the preparation of this book.

FENG GUO
FUYUHIKO TAMANOI

CHAPTER ONE

Dicer Proteins and Their Role in Gene Silencing Pathways

Michael Doyle*,1, Lukasz Jaskiewicz†, Witold Filipowicz‡,1

*Center for Brain Research, Medical University of Vienna, Vienna, Austria
†Biozentrum, University of Basel, Basel, Switzerland
‡Friedrich Miescher Institute for Biomedical Research, Basel, Switzerland
1Corresponding authors: e-mail address: michael.doyle@meduniwien.ac.at; witold.filipowicz@fmi.ch

Contents

Abstract

The RNase III family endoribonuclease Dicer is the key enzyme involved in the RNA interference (RNAi) and microRNA (miRNA) pathways. Apart from its function in biogenesis of miRNAs and short interfering RNAs (siRNAs), Dicer also plays a role in the assembly of the RNA-induced silencing complex (RISC), which functions as an effector complex in RNAi and miRNA repression. In this chapter, we review our current knowledge of Dicer from various organisms. We focus on the functions of individual domains of Dicer, how the enzyme recognizes and processes its substrates, and provide information on Dicer-associated proteins and their role in small RNA biogenesis and RISC assembly. We also discuss Dicer's role in gene and chromatin regulation processes in the cell nucleus.

The Enzymes, Volume 32
ISSN 1874-6047
http://dx.doi.org/10.1016/B978-0-12-404741-9.00001-5

1

1. INTRODUCTION

In the past 10 years, RNA interference (RNAi) and microRNA (miRNA)-mediated gene silencing pathways have emerged as key global regulatory mechanisms controlling gene expression in eukaryotes. Both pathways depend upon an enzymatic cascade that generates either 20- to 25-nt small interfering RNAs (siRNAs) or miRNAs both of which function by base pairing to target RNAs. Together with other proteins, including Argonautes, they form part of effector ribonucleoprotein (RNP) complexes that ultimately trigger translational repression and/or degradation of target RNAs resulting in changes in gene expression. Several dynamic classes of RNP complexes exist including si- or mi-RISCs (RNA-induced silencing complexes, acting posttranscriptionally) or RITS (RNA-induced transcriptional silencing complexes, acting at the chromatin level). Regardless of the type of RNP formed, all contain small RNAs derived from longer double-stranded RNA (dsRNA) or hairpin precursors that are processed to their shorter lengths by RNase III-type endonucleases, Drosha and Dicer. The genome-encoded miRNAs are transcribed as primary precursor miRNAs (pri-miRNAs) in the nucleus and most are initially cleaved by the Drosha/DGCR8 complex to generate precursor-miRNAs (pre-miRNAs). These are then exported via Exportin 5 to the cytoplasm where they are processed by Dicer to the double-stranded siRNA-like form of miRNA. Dicer is also responsible for generation of siRNAs, which are processed from longer dsRNAs accumulating in the cell as the result of either viral infection or antisense transcription, or are experimentally induced to specifically downregulate gene expression.

In 1998, Fire *et al.* [1] were the first to demonstrate that in *Caenorhabditis elegans* sequence-specific gene silencing can be induced by dsRNA. Soon afterward, it became apparent that long dsRNA is processed into siRNAs that act as the effectors of RNA silencing. After an intense search for the enzymatic activity responsible, Dicer was identified in *Drosophila* and demonstrated to process dsRNA to siRNAs [2]. Similar enzymes were soon described in other eukaryotes, including mammals, with a copy number and domain structure varying among the different organisms [3–5]. For example, four Dicer proteins are present in *Arabidopsis thaliana*, two in *Drosophila melanogaster*, and one in mammals, *C. elegans*, and *Schizosaccharomyces pombe*. Later, more "primitive" Dicer proteins were characterized from unicellular eukaryotes, including *Trypanosoma brucei*, *Giardia intestinalis*, and some budding yeast species [6,7].

In this chapter, we review our current knowledge about Dicer and Dicer-associated proteins. We will focus on the function of Dicer in RNA silencing and posttranscriptional regulation in various organisms as well as how structural and mutational studies have provided new insights in the way Dicer functions.

2. DICERS: RNASE III FAMILY MEMBERS WITH A ROLE IN GENE SILENCING

Dicer is a member of the RNase III family of endoribonucleases that specifically cleave dsRNA. All members of the family posses one or two copies of the RNase III domain which is responsible for recognizing and cleaving dsRNA. RNase III was first described in *Escherichia coli*, and since then, many endoribonucleases related to it have been identified that are important for RNA processing and posttranscriptional gene regulation in both prokaryotes and eukaryotes [8]. One of the most prominent roles of this class of proteins is their involvement in pre-ribosomal RNA (pre-rRNA) processing (reviewed by Nicholson [8]). Blaszczyk *et al.* [9] subdivided enzymes of the RNase III family into three classes, according to their domain organization, but in 2008, we proposed that RNase III orthologs be grouped into just two classes [3]. In the original classification, Drosha and Dicer proteins were grouped into separate classes due to the perceived complexity at the time regarding their respective domain organizations. Since the discovery of primitive Dicers, including those from *T. brucei*, *G. intestinalis*, and some budding yeast species (e.g., *Saccharomyces castelli*, *Kluyveromyces polysporus*, *Candida albicans*), having a less complex organization, the need to separate Dicer and Drosha proteins is not necessary and we continue to advocate that RNase III orthologs be grouped into just two classes. According to this classification, class I includes all bacterial and fungal proteins that contain a single RNase III domain as well as, generally, a C-terminal dsRNA-binding domain (dsRBD). This class of RNase III orthologs, functioning as homodimers, also contains some primitive Dicers, for example, *K. polysporus* Dicer which has a single RNase III domain and two dsRBDs. In contrast, class II RNase III orthologs function as monomers and contain the more complex Dicer and Drosha proteins with two tandem RNase III domains (termed RNase IIIa and IIIb) in the C-terminal region. In addition, Drosha and Dicer proteins contain multiple other domains in varying combinations and copies, including helicase/ATPase, PAZ (Piwi/Ago/Zwille), DUF283 (domain of unknown function), and dsRBD. Relevant

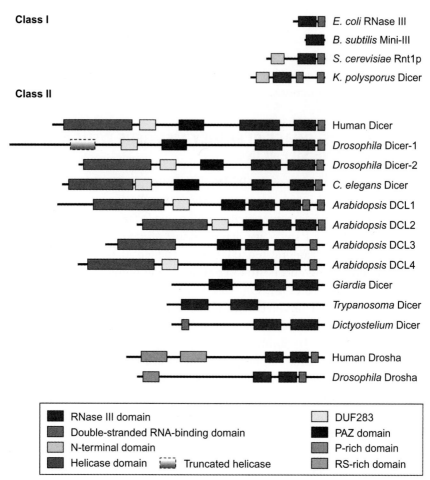

Figure 1.1 Classification of RNase III family enzymes into two classes. Class I contains bacterial and fungal RNase III orthologs as well as budding yeast Dicer, all containing a single RNase III domain. Class II contains Dicer and Drosha proteins containing two RNase III domains. Domain organization of the selected proteins of the RNase III family is shown schematically. (See color plate section in the back of the book.)

examples of proteins belonging to both classes of RNase III enzymes are schematically shown in Fig. 1.1.

The complexity of Dicer proteins can be viewed at multiple levels ranging from the various combinations of domains they contain in addition to the catalytic RNase III domains, the number of different proteins that are expressed in a single organism, to their localization and site of function in the cell. In animals, plants, and *S. pombe*, Dicers are large multidomain proteins. However, in most of lower eukaryotes, Dicer has a less complex domain organization. The simplest Dicer proteins studied to date are that found in *T. brucei* that contains just two RNase III domains and those active in some budding yeast species, containing single RNase III domain and dsRBD (Fig. 1.1). Vertebrates and also some other organisms (e.g., *C. elegans*, *S. pombe*) express only one Dicer protein. In insects and plants, which express multiple Dicer proteins, individual Dicers generally have specialized functions in gene silencing. Two Dicers proteins, termed Dcr-1 and Dcr-2, are expressed in *D. melanogaster*. Studies of mutant flies and biochemical experiments revealed that Dcr-1 is required for the processing of miRNA precursors, whereas Dcr-2 mainly processes siRNA precursors [10,11]. Delineating the individual functions of the four Dicer-like (DCL) proteins expressed in *A. thaliana* has proved more challenging. DCL1 functions in miRNA biogenesis at both pri-miRNA and pre-miRNA-processing steps as well as in siRNA production from endogenous inverted repeats. The other three DCLs are required for specific types of siRNA production: DCL2 processes natural *cis*-acting antisense transcripts into siRNAs and also functions in viral defense, DCL3 produces siRNAs which act as guides in chromatin modification, and DCL4 is the key enzyme generating *trans*-acting siRNAs (tasiRNA) which target mRNAs with only partially complementary sequences in a similar manner as miRNAs [12,13].

3. MECHANISM OF PROCESSING AND THE ROLE OF THE CATALYTIC RNASE III DOMAINS

Having identified Dicer to be a member of the RNase III family of endoribonucleases [2], a concerted effort began to understand: (a) how the enzyme cleaves its substrates to the products of defined length; (b) how substrate recognition is achieved; and (c) what is the role of individual domains in Dicer function. These questions were addressed biochemically, initially with the human Dicer, and were later complemented by structural studies, most notably those of *G. intestinalis*, human, and budding yeast Dicers. Human Dicer is a large protein of approximately 220 kDa that comprises, from the N- to C-terminus, a helicase/ATPase, DUF283 and PAZ domains, two RNase III domains, and a single dsRBD (Fig. 1.1). Early biochemical studies [14,15], performed with the purified human Dicer expressed in insect cells, revealed that the endonuclease activity of Dicer requires the presence of Mg^{2+} ions which can to an extent be substituted by Mn^{2+} or Co^{2+}. The enzyme was demonstrated capable of cleaving dsRNA substrates, ranging in size from 30 to 500 bp, to $\sim$20-bp siRNAs bearing 2-nt $3'$-overhangs and $5'$-phosphate and $3'$-OH termini, characteristic features of products generated by RNase III family enzymes [8]. The Dicer displayed a strong preference to progressively process dsRNA substrates starting from their ends. Blocking the ends of the substrate, either by capping them with stable tetraloops or by DNA–RNA duplexes, resulted in internal cleavages, but at much reduced levels [15].

The observation that generally Dicers are members of the class II of RNase III family focused attention on the role that the two RNase III domains play in the catalysis. The class I RNase III enzymes function as homodimers and structural characterization of the bacterium *Aquifex aeolicus* (Aa) RNase III and the mutagenic analysis of the *E. coli* RNase III (Ec-RNase III) [9], suggested a model proposing that RNase III dimer contains two catalytic centers, each cleaving both strands of the dsRNA substrate at nearby phosphodiester bonds [9]. The spacing of the catalytic centers on the dsRNA predicted generation of 9-bp products with 2-nt $3'$-overhangs, consistent with the size of products generated by bacterial RNase III *in vitro* [8,9].

Given that most Dicer enzymes contain two tandemly arranged RNase III domains and produce longer products, immediately indicated that the mechanism of processing must be different from that proposed for prokaryotic enzymes. Sequence alignments of the human Dicer RNase IIIa and IIIb domains with bacterial enzymes indicated the presence of conserved acidic residues likely functioning in dsRNA cleavage and Mg^{2+} coordination in the human enzyme. Mutation of these residues in context of the full-length human protein (D1320 and E1564 from RNase IIIa; and D1709 and E1813 from IIIb; these residues are equivalent to the Ec-RNase III residues D44 and E110) demonstrated that they, but not additional other residues proposed as also catalytic in the bacterial model [9], are essential for dsRNA cleavage [16]. Analysis of the cleavage products generated by catalytic residues mutants in either the RNase IIIa or IIIb domain revealed that Dicer accesses its substrates in a polar fashion. Whereas the RNase IIIa domain always processes the protruding $3'$-OH-bearing RNA strand, the RNase IIIb cuts the opposite recessed $5'$-phosphate-containing strand. In pre-miRNA, these stands correspond to the descending arm of the hairpin cut by domain IIIa and the ascending arm cut by IIIb. Taken together with the results of gradient sedimentation and modeling of the Dicer RNase III domains in a complex with dsRNA, these data led to a conclusion that Dicer functions as an intramolecular pseudo-dimer with the RNase IIIa and IIIb

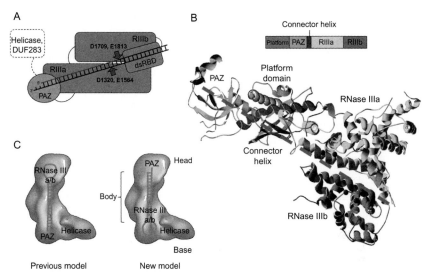

Figure 1.2 Structural organization of Dicer. (A) A mutagenesis-based model of dsRNA processing by human Dicer [16]. Individual domains of Dicer are shown in different colors. The enzyme contains a single dsRNA cleavage center with two independent catalytic sites. The catalytic center is formed by intramolecular dimerization of the RNase IIIa and RNase IIIb domains. Most important catalytic residues are indicated. DsRBD positioning is arbitrary. (B) Crystal structure of *G. intestinalis* Dicer [7]. This ribbon representation of Dicer shows the N-terminal platform domain (blue), the PAZ domain (orange), the connector helix (red), the RNase IIIa domain (yellow), and the RNase IIIb domain (green). Positions of metal ions are indicated by violet dots. Predicted location of dsRNA is indicated by a dashed line. PDB_id 2ffl was rendered using DeepView [17] and POV-Ray 3.6 (www.povray.org). (C) EM reconstruction of the shape of a full-length human Dicer. Left panel shows the originally proposed domain arrangement along with bound pre-miRNA [18–20]. Right panel shows the revised domain positioning based on streptavidin labeling of inserted tags [21]. Positions of individual Dicer domains and bound pre-miRNA, in relation to the head, body, and base regions of the letter L-shaped protein, are indicated and see Ref. [21]. *Figure (C), Right panel: Adapted from Ref. [20].* (See color plate section in the back of the book.)

domains forming a single-processing center containing two independent catalytic "half sites." Each site is capable of cutting one strand of the dsRNA duplex generating products with a 2-nt 3′-overhangs (Fig. 1.2A). Importantly, mutational analysis of Ec-RNase III and processing of its substrates indicated that the bacterial enzyme also uses the single dsRNA-processing center mechanism similar to that described for Dicer [16].

Validity of the proposed catalytic models for both Dicer and prokaryotic RNase III were soon confirmed by additional biochemical and also crystallographic studies. Most revealing was solution of the crystal structure of a full-length Dicer protein from the protozoan *G. intestinalis* [7]. This "primitive" Dicer is smaller than its orthologs in higher organisms and only contains DUF293-like, PAZ and RNase III domains. Structural analysis revealed it forms an elongated molecule that may act as a molecular ruler measuring the distance from the end of the dsRNA substrate, recognized by the PAZ domain, and the sites of enzymatic cleavage. The intramolecular dimer of the RNase IIIa and b domains forms the catalytic core while the PAZ domain is linked to it via a conserved "connector" α helix that may determine overall product size (Fig. 1.2B). The distance between the PAZ and the active sites of RNase III domains was calculated as ~65 Å which matches the length of ~25 bp, the typical size of siRNA produced by *Giardia* enzyme [7]. The connector α helix contains an evolutionarily conserved proline residue, which induces a distinct kink that may help different Dicers to accommodate helix irregularities found, for example, in many pre-miRNAs (Fig. 1.2B).

In both human [16] and *Giardia* [7] studies, the dsRNA substrate was modeled to occupy the RNase III catalytic "valley" which contains many conserved amino acids. Mutation of many positively charged residues in the valley impaired activity of the *Giardia* Dicer, supporting their role in substrate binding [22]. Gan *et al.* [23] were first to succeed in solving the structure of the RNase III family enzyme, Aa-RNase III, in a catalytic complex with dsRNA. The structure revealed dsRNA placement similar to that predicted by modeling but also identified several protein motifs responsible for the substrate binding. Crystallographic studies discussed above, followed by additional structural work on both Aa-RNase III [24] and the budding yeast Dicer [25] (see below) also identified all amino acid residues contributing to the catalysis by RNase III enzymes. Consistent with previous indications stemming from both structural [7,24] and biochemical [26], the RNase III enzymes use the two-Mg^{2+}-ion catalytic mechanism for hydrolysis of each phosphodiester bond. Of the two residues absolutely essential for the catalysis, E110 (Aa-RNase III numbering) coordinates with both Mg^{2+} ions (MgA and MgB), while D44 coordinates with MgB. Still other amino acids provide required coordination during catalysis [24].

Interestingly, despite that Dicers generally function as intramolecular pseudo-dimers of RNase IIIa and IIIb domains, the isolated RNase IIIb domain of both mouse and human Dicers can also assemble into a catalytically active homodimer similar to that of the prokaryotic RNase III [27,28]. Structural analyses of the isolated human RNase IIIb domain alone [28] or the mouse RNase IIIb-dsRBD fragment in complex with dsRNA [27] provided further insights into the nature of Dicer's catalytic core. Analysis of the isolated human RNase IIIb domain identified two Mg^{2+} ions per monomer around the active site. The location of the MgA ions was similar to position of ions identified in *Giardia* Dicer and Aa-RNase III, supporting a conserved mechanism of catalysis. However, the location of the MgB ions was not identical, raising the possibility that they may also be involved in coordinating RNA binding [28]. The crystal structure of mouse Dicer fragment in complex with dsRNA revealed an additional conserved lysine residue critical for dsRNA cleavage [27]; a corresponding residue is also important for activity of *K. polysporus* Dicer [25] (see below). Both RNase IIIb domain studies also identified a metazoan-Dicer-specific ∼45-aa flexible loop bearing an extra α helix, providing potential of the domain to engage with the minor groove of dsRNA. Interestingly, the RNase IIIa domain also contains a conserved flexible loop (referred to as a positioning loop; residues 391–401 in *Giardia* Dicer), which is important for dsRNA binding and its proper positioning at the enzyme active site [22]. The work of Du *et al.* [27] also provided the first structural determination of a Dicer dsRBD, confirming that in mammals it possesses the canonical α−β−β−β−α−fold. It will be interesting to find out whether the RNase IIIb homodimeric interaction also occurs *in vivo*, between two Dicer molecules or their variants.

Initial data on the polarity of pre-miRNA cleavage [16,29] are nicely complemented by a mutagenic study where various versions of full-length Dicer mutated at key residues were analyzed in Dicer knockout murine fibroblasts [30]. This work revealed that mutations of individual RNase III domains have specific effects on processing of different pre-miRNAs. Inactivation of the IIIa domain resulted in the complete loss of miRNAs derived from the 3′-descending arm of the pre-miRNA. Conversely, miRNAs originating from the 5′-ascending pre-miRNA arm were lost when the IIIb domain was inactivated. Finally, the pre-miRNA processing was partially affected by some mutations of the PAZ domain, thus underscoring the importance of this domain in biogenesis of miRNAs [30]. A specific role of the RNase IIIa domain in processing the descending arm of miRNAs was also identified in a genetic screen performed in Chinese hamster ovary (CHO) cells [31]. Additionally, mutations in the catalytic core of the RNase IIIb domain that impair miRNA biogenesis have been identified from patients with nonepithelial ovarian tumors [32].

4. ROLE OF OTHER DOMAINS IN DICER FUNCTION

4.1. PAZ domain

Of all the noncatalytic domains in Dicer, the PAZ domain has been one of the most intensively studied not least because it is also found in Ago proteins where it was first demonstrated to recognize the 3′-protruding nucleotides of an siRNA end (see Chapter 4). Structurally, the PAZ domain resembles the oligonucleotide/oligosaccharide-binding (OB) fold. Mutations of the key PAZ residues implicated in RNA binding (F960, YY971/972, and E1036) in the context of the full-length human Dicer resulted in reduced substrate processing [16]. Consistent with this observation was the finding that Dicer cleaves dsRNA and pre-miRNA substrates containing 3′-overhangs more efficiently than those bearing blunt ends. This preference was accentuated when Dicer mutants devoid of the dsRBD were used. These data indicated that the PAZ domain recognizes the 3′-overhang-containing end of substrate RNA, similar to its role in Ago proteins. Function of the PAZ domain in recognizing the end of dsRNA and in determination of the length of cleavage products also emerged from X-ray structure of *G. intestinalis* Dicer [7] (see above). This was further confirmed when mutational analysis identified a conserved loop in the Dicer PAZ domain whose structural position is important for both dsRNA end recognition and generating products of the correct length [22]. Replacement of the PAZ domain in the context of full-length Dicer with the RNA-binding domain of the spliceosomal protein U1A resulted in the protein gaining specificity for dsRNA substrates bearing the U1A recognition loop and yielding products of different length than the wild-type enzyme [22].

Recently, structural and biochemical studies provided new information on role of the PAZ domain in human Dicer [33]. The domain was demonstrated to anchor not only the 3′-end of the substrate but also its phosphorylated 5′-end, with the position of cleavage being determined by the ~22-nt distance from the 5′-end. This has since been termed the 5′-counting rule. Mutational analysis of the human Dicer PAZ domain identified a novel basic motif termed the 5′-pocket which is required for the recognition of the 5′-phosphorylated end of the dsRNA substrate. If mutated, the *in vitro* efficiency of substrate processing is reduced and the position of cleavage sites altered thus underscoring the importance of 5′-pocket for precise and effective processing of substrates by Dicer. The 5′-counting rule and 5′-pocket are conserved in *D. melanogaster* Dcr-1 but not *G. intestinalis* Dicer [33].

4.2. Helicase/ATPase

Most of the canonical Dicers contain at N-terminus a large helicase domain which shares homology to the DExD/H-box family of helicase proteins that play multiple roles in RNA metabolism, including the ATP-dependent dissociation of RNA duplexes or translocation along RNA. Early biochemical studies, however, did not provide any direct mechanistic insight into the role of this domain within Dicer: human Dicer was shown not to require ATP to process its substrates and mutation of the conserved Walker A motif (this mutation strongly inhibits nucleotide binding and enzymatic activity in helicases/ATPases and other P-loop-containing proteins) had no observable effect on enzymatic function [14,15]. In contrast, the presence of the helicase domain was found to correlate with the observed requirement of ATP for siRNA production in invertebrate Dicers [2,34] (and see below).

Recently, new evidence for the function of the helicase domain has emerged for both vertebrate and invertebrate Dicers. In human Dicer, it has been suggested to play multiple roles in the enzymes function. Deletion of the helicase domain resulted in the activation of the enzyme with the efficiency of processing increasing depending on the substrate assayed [35]. For a 37-bp perfect duplex-substrate, the processing increased by up to 65-fold whereas there was only a modest stimulation of processing of human pre-let-7 miRNA, indicating the domain may contribute to distinguishing between perfect dsRNA substrates and those represented by stem-loop hairpins. Such a role has since been confirmed when it was demonstrated that (i) the isolated helicase domain preferentially binds the pre-let-7 miRNA with higher affinity compared to the 37-bp perfect duplex-substrate [35]; (ii) the addition of a terminal loop to the 37-bp perfect duplex-substrate increases the rate of processing by full-length Dicer to the levels observed for the pre-let-7 miRNA [36]; and (iii) defects in miRNA processing were observed in a human Dicer mutant cell line where the helicase domain is disrupted by a 43-amino acid insertion [37]. Additionally, kinetic analysis revealed that upon deletion of the helicase domain, the observed increase in processing was due to enhanced catalysis and not an increase in substrate binding. This suggests that the helicase domain interferes with the functionality of the catalytic domain and maintains Dicer in a semi-repressed state [35]. Interestingly TRBP, a dsRBD-containing partner of Dicer (see below), interacts with the helicase domain and stimulates full-length Dicer activity, suggesting that protein cofactors of Dicer may modulate the helicase domain-mediated inhibition. Limited proteolysis has also been demonstrated to enhance activity of human Dicer, suggesting that structural rearrangements, possibly involving the helicase domain, also contribute to enhancing enzymatic activity [15,38].

In contrast to the human enzyme, the helicase domain in the *C. elegans* Dicer or *D. melanogaster* Dcr-2 is not necessary for miRNA processing but is instead essential for the ATP-dependent production of many endogenous siRNAs (endo-siRNAs) [10,39,40]. Consistently, mutations in the Walker A or DExD motifs of Dcr-2 reduce siRNA production [10,40]. The *in vitro* processing assays using either *C. elegans* extracts or purified *D. melanogaster* Dcr-2, with and without helicase domain mutations, demonstrated that the domain contributes to substrate recognition [39]. Moreover, evidence was obtained that *Drosophila* Dcr-2 is a dsRNA-dependent ATPase and that ATP hydrolysis converts it from a distributive to a processive enzyme [10]. Similarly, the generation of siRNAs by the purified *S. pombe* Dicer is dependent on ATP and the integrity of the Walker A motif, and *S. pombe* strains expressing Dicer with a Walker A mutation are defective for centromeric silencing and formation of siRNAs [41].

Most recently, the similarity of Dicer's helicase domain to retinoic acid-inducible gene-I (RIG-I) has been explored both at the structural and *in silico* levels. RIG-I and related receptors are involved in antiviral immune response through their recognition of viral RNAs, and they contain a helicase domain closely related to that of Dicer. Specifically, the helicase domain in RIG-I-like proteins and Dicers contains a long insertion, HEL2i, in the second RecA-like fold (HEL2) of the domain. Structural studies of RIG-I indicated that upon dsRNA binding, the helicase domain rearranges to clamp down on its substrate, with the HEL2i insertion greatly contributing to dsRNA binding [42–44]. A similar role has also been suggested for the helicase domain of human Dicer [21] (see below).

4.3. DUF283

The DUF283 domain is conserved between Dicer orthologs and is not found in other proteins. Sequence and structure analyses suggest that it represents a divergent dsRBD loosely resembling the canonical dsRBD [45,46]. Although the role of DUF283 remains poorly understood, several studies have provided experimental data on its potential contribution to Dicer function. In human Dicer, deletion of DUF283 resulted in a reduction in substrate cleavage independent of any ability it may have to bind dsRNA [35,47]. Likewise, deletion of DUF283 prevented pri-miRNA processing by the recombinant *Drosophila* Dcr-1 [29]. Furthermore, DUF283 has been suggested to act as a protein–protein interaction domain in both DCL1 and DCL4 of *Arabidopsis* [46].

4.4. dsRBD

DsRBDs are conserved motifs found in a wide variety of proteins interacting with dsRNA. Multiple structures of individual dsRBDs together with dsRNA have been solved, all revealing a common $\alpha-\beta-\beta-\beta-\alpha$ topology where the two α helicases lay on the face of three antiparallel β sheets. All contacts between the domain are with the RNAs sugar–phosphate backbone, thus indicating that dsRBDs possess no primary sequence specificity [48,49]. The structure of the mouse Dicer dsRBD has been solved revealing it possesses the canonical $\alpha-\beta-\beta-\beta-\alpha$–fold [27]. Dicers generally possess either one or two copies of a dsRBD at the C-terminus. However, unlike other noncatalytic domains, the dsRBD is also present in class I proteins thus implying it plays an important role in activity of RNase III enzymes. The dsRBD is not essential for catalysis by either class of enzymes. For example, when the dsRBD of Ec-RNase III is deleted in the context of the full-length protein, the truncated enzyme continues to accurately process substrates *in vitro* [50]. Similarly, the dsRBD is dispensable for substrate processing by human Dicer: early biochemical studies demonstrated that deletion of the dsRBD in the context of the full-length enzyme only reduced the efficiency of substrate processing *in vitro* [16]. In contrast, human Drosha requires the domain for its activity in the miRNA pathway [51]. Therefore, although the dsRBD is found in the majority of RNase III enzymes, its presence is not always a prerequisite for the processing of substrates. This conclusion is underscored by the observation that some primitive Dicers, including those from *G. intestinalis* and *T. brucei*, do not contain dsRBDs but are still effective in processing substrates capable of triggering RNAi [7,52]. Consequently, it is thought that, during processing, dsRBDs stabilize the interaction of Dicer with the substrate. Such a role is supported by structural data of Aa-RNase III, which indicate that two dsRBDs of the dimer play an important role in binding of dsRNA by securing it from the top into the catalytic core and dynamically moving up and down the minor groove of the substrate during catalysis [23,24]. Recent data for human Dicer show that the dsRBD is only required for dsRNA binding in the absence of the PAZ domain and plays no role in determining the length of cleavage products [36], consistent with previous observations [16]. Taken together, the data on dsRBD function point to it having only an auxiliary role in dsRNA binding and cleavage by Dicer. However, the role of the dsRBD in Dicer has recently been extended beyond its function in dsRNA binding as the dsRBD has been shown to modulate the nuclear-cytoplasmic distribution of the enzyme in *S. pombe* [53] (see below).

5. DOMAIN ARRANGEMENT IN METAZOAN DICERS

The studies described above, providing insight to the role of individual Dicer domains, focused mainly on the biochemical and structural analysis of isolated domains or subregions of the enzyme. The only structure available to date of a full-length class II Dicer protein remains that from *G. intestinalis*, containing platform, PAZ and RNase III domains and hence being significantly smaller and less complex than its orthologs in higher organisms (Fig. 1.2B). Due to their complexity and size, metazoan Dicers have proved recalcitrant to crystallization, and therefore, different approaches have been taken to understand the overall architecture of the enzyme.

Several studies have reported single-particle negative-stain electron microscopy (EM) reconstructions of human Dicer, all of which report an "L"-shaped molecule comprising several morphologically discrete regions [18,19,21] (Fig. 1.2C). However, due to the complexity of localizing individual domains on EM maps, especially to relatively small asymmetric particles such as Dicer, these first models of human Dicer architecture were conflicting. In one model, both the PAZ and helicase domains were placed in the main body (vertical long "L" arm) of the molecule toward its base (horizontal short "L" arm) with the RNase III domains above them at the top [19]. However, placement of the domains in this orientation is difficult to reconcile with biochemical data pointing to the role of the helicase domain in recognition of the loop of pre-miRNA substrates and in modulating activity of RNase III domains. The second model suggested two possible architectures for human Dicer, with the catalytic core at opposite ends of the molecule [18]. Additionally, the resolution of the EM data for this model was insufficient to accurately determine the position of the helicase domain. Consequently, although both models confirmed Dicer to be an L-shaped molecule with morphological discrete regions, the relative position of individual domains remained unclear.

To overcome the issue of defining the precise position of individual domains, the latest EM model of human Dicer made use of a site-specific tagging strategy compatible with single-particle analysis [21]. This allowed the 3D positions of the PAZ, RNase III and helicase domains to be experimentally determined. Again an L-shaped molecule was observed but the overall arrangement of domains was different compared to the initial models. In the new model, the PAZ and the RNase III domains are separated from each other, with the RNase III domains at the base of the molecule and the PAZ domain at the top (head) of the body. As with the first model, the helicase domain was placed at the base of the molecule in close proximity to the RNase III catalytic core, thus accommodating the biochemical data that it regulates processivity and substrate binding (Fig. 1.2C). Positioning of the helicase domain in the base of the molecule is further supported by 3D reconstructions of a Dicer devoid of the domain. Analysis of the helicase mutant revealed an oblong structure with dimensions similar to the vertical "L" arm of the full-length Dicer molecule, indicating the helicase domain forms the entire base [21]. Comparing the model of Dicer to the available crystal structures of RIG-I helicase [42–44] identified two discrete conformations of the Dicer helicase domain that resemble the apo- and dsRNA-bound crystal structures of RIG-I. For RIG-I, these structures demonstrated that upon dsRNA binding, the helicase domain clamps down on dsRNA substrate; the Dicer helicase domain may play a similar role in substrate binding [21]. The Dicer model of Lau *et al.* [21] also proposes that catalytic core is rotated relative to the PAZ domain when compared to the structure of *Giardia* Dicer. As a result, the distance between the catalytic core and the PAZ domain is shorter thus accounting for the differences in size of siRNAs generated by the two Dicers. EM analysis of *D. melanogaster* Dcr-2 indicated that it also forms L-shaped molecules, suggesting conservation of Dicer architecture [21].

Reconstitution of active forms of human Dicer from its subfragments expressed in *E. coli* further enhanced understanding of how the individual Dicer domains cooperate to faithfully process its substrates. These experiments revealed that each RNA-interacting domain is employed in distinct aspects of the dicing reaction and in the recognition of different substrate classes [36]. Three distinct modes of recognition were determined: (a) anchoring by the PAZ domain of the end of a dsRNA helix followed by its interaction with the RNase III domains to reconstitute an accurate but nonsubstrate-selective dicing complex; (b) in the absence of the PAZ domain nonspecific RNA binding by the dsRBD becomes important for substrate recruitment, consistent with earlier indications that the dsRBD enhances the stability of the complex [16]; (c) selection of substrate classes by the helicase domain via its specific interaction with the pre-miRNA hairpin loop, absent in dsRNA substrates [36]. A similar role for the helicase domain in distinguishing pre-miRNAs from other substrates has also been described for *D. melanogaster* Dcr-1. Although fly Dcr-1 contains a defective helicase domain, the domain still recognizes the single-stranded pre-miRNA terminal loop, checks the loop size, and also contributes to PAZ domain measurement of the distance between the 3'-overhang and terminal loop [54]. This allows Dcr-1 to strictly inspect the authenticity of pre-miRNAs substrates and helps explain why it exclusively cleaves pre-miRNAs. On the other hand, Dcr-2 which contains a functional helicase/ATPase domain has evolved to process long dsRNAs in ATP-dependent way, with inorganic phosphate and R2D2 partner protein further helping the enzyme to discriminate against processing of pre-miRNAs [10].

Taken together with the early biochemical and structural data, the experiments discussed above give us a comprehensive overview of how a complex canonical Dicer functions. The enzyme is an L-shaped molecule, with RNase III and helicase domains at the base and the PAZ domain at the top of the body. In this organization, the molecular ruler between the RNase III and PAZ domains, comprised of a connector α helix and the platform domain, determines the size of the cleavage product while at the same time allowing the biochemical observations of RNA binding by the helicase domain to be accommodated. Furthermore, positioning the helicase domain next to the RNase III domains, supported by both EM [21] and biochemical [54] data, helps to reconcile the distinct function the domain plays in different Dicer homologs. For instance, its position allows it to bind both the dsRNA and loop segments of pre-miRNA substrates as demonstrated for human Dicer and *D. melanogaster* Dcr-1. For processive Dicers that bind and cleave long dsRNAs sequentially into siRNAs, the domain is well positioned to translocate dsRNA into the catalytic core via ATP hydrolysis and to remain bound to long dsRNA substrate after cleaving away single siRNA products. This supports the biochemical observations from *D. melanogaster* Dcr-2, demonstrating that the helicase domain is essential for the generation of multiple siRNAs from a dsRNA before it dissociates from the substrate [10]. Thus, this model provides a structural basis of how the helicase domain could contribute to the processive cleavage of long dsRNA substrates. Finally, the model is supported by the recent reconstitution and mutagenic studies, which also demonstrate that individual domains contribute both specifically and dynamically to the recognition of different substrates.

6. INSIDE-OUT PROCESSING MECHANISM BY THE YEAST DICER

Until recently, studies on Dicer processing primarily focused on more complex canonical Dicers possessing two RNase III domains, a PAZ domain, and various other domains. These Dicers recognize the end of dsRNA substrate through the PAZ domain, sequentially cleaving it during multiple rounds of processing, generating products of the correct length. They use similar PAZ-dependent mechanism to process pre-miRNAs. As they process from the end of the substrate inward, this type of cleavage can be referred to as the "outside-in" mechanism of processing. The discovery of non-canonical Dicers possessing only a single RNase III domain and no PAZ domain [6,25] immediately raised the question of how these enzymes recognize, bind to, and cleave their substrates at precise intervals *in vivo*.

Dicer of the budding yeast *K. polysporus* or *S. castellii* possesses a single RNase III domain, two dsRBDs, and a unique N-terminal domain (NTD) [6]. *In vitro* assays using recombinant *K. polysporus* Dicer demonstrated that it successfully processes a substrate dsRNA into 23-nt siRNA products thus arguing against a protein cofactor that substitutes *in vivo* for the absence of a PAZ domain [25]. Deletion analysis identified a core region comprising the RNase III and first dsRBD that is capable of generating siRNA products indistinguishable from those of the full-length enzyme. The solving of the crystal structure of this region, and also fragment encompassing NTD, RNase III domain, and dsRBD1, confirmed that like other class I RNase III enzymes, *K. polysporus* Dicer acts as a homodimer. The NTD was found to also greatly contribute to the dimerization. Interestingly, the NTD shows sequence similarity with the N-terminal region of the yeast RNase III Rnt1, suggesting a common evolutionary origin of both RNase III family enzymes. The structure also identified the core catalytic residues within the RNase III domain as well as two additional residues contributing to metal-ion binding and conserved throughout eukaryotic RNase III enzymes [25].

Unlike canonical Dicers that recognize and bind to the end of the substrate and process from the outside in, the *K. polysporus* Dicer initially binds as a dimer centrally within the substrate. This is followed by the recruitment of additional dimers to adjacent regions in dsRNA, resulting in precise alignment of enzymes molecules so when cleavage occurs, the 21-bp siRNAs with 2-nt 3'-overhangs are generated. Analysis of the turnover kinetics of the cleavage reaction revealed that product release is slow relative to substrate binding and/or catalysis. This is essential as this model relies on a second dimer binding cooperatively at the proper position before the first dimer cleaves and releases its product. In addition, slow product release may protect siRNAs from further cleavage as well as facilitate the involvement of Dicer in RISC loading similar to that reported for canonical Dicers. As initial binding occurs inside the substrate dsRNA, this mode of cleavage has been termed the "inside-out" mechanism and in distinction from the "outside-in" mechanism of canonical Dicers [25] (Fig. 1.3).

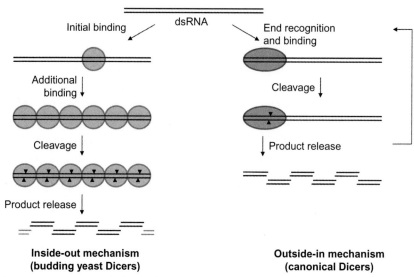

Figure 1.3 Inside-out and outside-in mechanisms of dsRNA processing by Dicer. Two mechanisms of Dicer cleavage are shown. In the inside-out mechanism, the budding yeast Dicer first binds internally to dsRNA substrate, followed by cooperative recruitment of further molecules. Alignment results in cleavage and generation of products of correct length. In the outside-in mechanism of canonical Dicers, the enzyme always binds the end of the substrate, progressively cleaving away individual siRNA products. *Figure modified from Ref. [25].*

7. FUNCTION OF DICER IN THE NUCLEUS

In the generic model of RNAi, Dicer functions in the cytoplasm where it cleaves its substrates before they are loaded into the Ago–containing RISC, ultimately leading to gene silencing. In recent years, the role of Dicer in the nucleus has been receiving much attention especially in plants and invertebrates where Dicer has been shown to localize to the nucleus and to play a role in transcriptional gene silencing. Limited evidence exists that Dicer may also localize to and be functional in the nucleus in mammals.

The best studied example of Dicer nuclear function is that from *S. pombe* where the enzyme Dcr1 was shown to actively participate in transcriptional and cotranscriptional gene silencing (TGS and CTGS) [55,56]. These pathways function in association with chromatin and can trigger the formation of heterochromatin at centromeric repeats and the silent mating-type loci [57,58]. A requirement for RNAi proteins in the assembly of heterochromatin in *S. pombe* was discovered a decade ago [55].

In this organism, deletion of genes coding for Dcr1, Ago, or RNA-dependent RNA polymerase results in nuclear phenotypes, including the loss of TGS/CTGS, chromosome segregation defects, and reduced levels of repressive histone marks [57–59]. More recently, Dcr1 has been demonstrated to also function outside constitutive heterochromatin. Using DNA adenine methyltransferase identification (DamID), it was demonstrated that Dicer physically associates with some euchromatic genes, noncoding RNA genes, and retrotransposon long terminal repeats. This physical association of Dicer with chromatin is sufficient to trigger a silencing response but not the assembly of heterochromatin [56].

S. pombe Dcr1 is a canonical enzyme with two RNase III domains, a PAZ-like, DUF293, and helicase domains, and a dsRBD at the C-terminus (Fig. 1.1). It processes siRNAs that correspond to centromeric repeat sequences and that are loaded into an Ago-containing RITS complex. RITS complexes associate with chromatin through base pairing between siRNAs and the nascent RNA polymerase II (Pol II) transcripts, as well as with H3K9 methylated nucleosomes through the chromodomain of the Chp1 subunit. RITS can recruit histone-modifying enzymes (e.g., Clr4) [60], which is crucial to assemble and maintain silent heterochromatin at centromeric repeats [57–59].

Additionally, Dcr1 has been implicated in regulating the termination of transcription at convergent genes (CGs), representing pairs of adjacent genes expressed from opposite strands. Due to inefficient termination, their transcription by Pol II during G1–S cell-cycle phase may lead to the formation of overlapping read-through transcripts, creating dsRNA substrates for Dcr1. Consequently, transient, RITS-mediated methylation of H3K9 at CGs in G1–S would then enable the recruitment of cohesin, which would remain bound to the CG throughout G2 and prevent further transcriptional read-through [61]. Interestingly, many genes-encoding RNAi components including Dcr1 are themselves convergent, resulting in their cell-cycle-dependent autoregulation, via a similar mechanism as mentioned above [62].

Recent studies of S. pombe Dcr1 have revealed new functions of the enzyme and its C-terminal dsRBD, with potential implications for other Dicers. Live-cell imaging of Dcr1-GFP revealed predominant localization of the enzyme in the nucleus, enriched in foci colocalizing with nuclear pores on the inner side of the nuclear membrane [53]. Deletion experiments revealed that the dsRBD plays a crucial role in subcellular distribution of Dcr1. Deletion of the dsRBD or just the very C-terminal 33 amino acids (Dcr1ΔC33) resulted in predominantly cytoplasmic localization of the enzyme. It was further demonstrated that Dcr1ΔC33 actively shuttles between the nucleus and the cytoplasm. Importantly, rather than acting as a nuclear localization signal, the C33 fragment was found to block dsRBD-mediated nuclear export, resulting in nuclear retention of Dcr1 at pores. Subsequent structural studies revealed that C33 contributes to the proper folding of Dcr1's dsRBD. It was found that the C-terminus of *S. pombe* Dcr1 adopts a typical dsRBD fold, but additional extended structural elements including a novel zinc-finger motif that is conserved among fission yeast Dicers are also present [63]. Zinc coordination by this motif is required for nuclear retention of Dcr1 and depends on residues that are encoded in both the dsRBD and C33, explaining the observed phenotypes of Dcr1ΔC33. These are loss of siRNA production and heterochromatin silencing [53], as well as impaired repression of stress-response genes [64]. Interestingly, similar to disrupting zinc-coordination genetically, Dicer is deported to the cytoplasm under chronic heat shock conditions, most likely due to temperature-induced structural changes in its C-terminus. Thus, the dsRBD of Dcr1 has been proposed to constitute a thermoswitch that controls the conditional relocalization of Dicer to the cytoplasm, providing a mechanism for the release of stress-response genes from RNAi repression during chronic heat shock [64]. In summary, these studies demonstrated an important role of the Dcr1 extended dsRBD in the nucleo-cytoplasmic shuttling and specific subnuclear localization of the enzyme.

Nuclear roles for Dicer are also established in *Drosophila* and plants. In *Drosophila*, it was observed that Dcr-2 along with other RNAi components frequently colocalizes with polycomb group (PcG) proteins on chromosomes [65]. PcG complexes bind to polycomb response elements (PREs) at or near the promoters of homeotic genes leading to transcriptional repression. PREs are present in multiple copies in the genome and they cluster together resulting in enhanced silencing. This is defective in Dcr-2 as well as other RNAi mutants, suggesting that Dcr-2 contributes to nuclear organization in *Drosophila*. Furthermore, Dcr-2-dependent small 21–23-nt RNAs, derived from sense and antisense transcripts from specific PREs, are detectable [65]. More recently, Dcr-2 was also found preferentially associated with transcriptionally active euchromatic regions and shown to interact with the core transcription machinery including Pol II [66]. Dcr-2 was found associated with sense and antisense transcripts and its loss perturbed the position of Pol II on promoters effecting transcription, particularly during heat shock [66].

In plants, complex nuclear roles for all four Dicers have been established (for review, see Refs. [12,13,67–69]). miRNA biogenesis specifically depends on DCL-1 which is responsible for the processing of pri-miRNAs in the nucleus [70,71]. In terms of direct nuclear gene silencing, RNA-directed DNA methylation (RdDM) was the first RNA-guided epigenetic modification to be discovered [72] and was subsequently shown to require dsRNA processed into small 21- to 24-nt RNAs in both tobacco and *Arabidopsis*, thus suggesting a role for Dicer function [73]. DCL3 is thought to be the most critical for this process and has been shown to localize to the nucleus and be responsible for producing siRNAs capable of triggering RdDM [74,75]. Additionally, DCL2 is also implicated in RdDM [75] (for a recent review on RdDM, see Ref. [69]). Recently, a cotranscriptional role for DCL4 was found in regulating transcription termination of an endogenous *Arabidopsis* gene [76].

In mammals, direct nuclear roles for Dicer remain elusive. Mammalian Dicers do not possess extended dsRBDs with zinc-finger motifs shown to mediate nucleo-cytoplasmic shuttling of the enzyme in *S. pombe* [53]. However, the ectopically expressed human enzyme also shuttles between the nucleus and the cytoplasm and the dsRBD plays a role of this process (our unpublished results). The presence of Dicer in the nucleus in human cells is also supported by the finding that it can associate with the nuclear pore complex protein NUP153 [77]. Dicer-dependent processing of small nucleolar RNAs (snoRNAs) into miRNA-like molecules has been described for mammalian cells, further supporting the possibility that Dicer may be active in the nucleus [78,79]. Also consistent with the nuclear function, Dicer was found to be associated with ribosomal DNA (rDNA) repeats in human and mouse cells [80]. Biological significance of this observation is not clear since the enzyme appears to be associated with both non-CpG-methylated active and methylated silent rDNA genes. Moreover, Dicer knockout had no significant effect on chromatin status at rDNA loci or rRNA biogenesis in mouse embryonic stem (ES) cells [80].

Other studies of Dicer-deficient mouse ES cells, however, did note some nuclear phenotypes, potentially due to the loss of the enzyme. These include reduced epigenetic silencing of centromeric repeats and increased telomere recombination and elongation in Dicer-deficient mouse ES cells [81–83]. Studies of Dicer-deficient ES cells have also implicated the enzyme as having a role in X-chromosome inactivation although this remains controversial [84–86]. In the chicken-human hybrid DT40 cells, loss of Dicer leads to premature sister chromatid separation due to abnormalities in heterochromatin formation [87]. Additionally, in mouse oocytes having Dicer conditionally deleted, phenotypes including multiple disorganized spindles and severe chromosome congression defects were observed [88]. Dicer has also been implicated in regulating intergenic transcription at the human β-globin locus [89]. However, the aforementioned studies do not exclude the possibility that the observed phenotypes are an indirect consequence of a failure of miRNA biogenesis in the cytoplasm rather than a loss of Dicer from the nucleus. For example, Dicer deletion affects DNA methylation in mouse ES cells but this is at least partially due to miRNAs controlling expression of the *de novo* DNA methyltransferase Dnmt3 [90]. In summary, although there are many observations suggesting the Dicer role in vertebrate nucleus, more direct evidence is required to demonstrate its function in this compartment.

8. PROTEINS INTERACTING WITH DICER AND THE ASSEMBLY OF RISC EFFECTOR COMPLEXES

Although biochemical analyses revealed that recombinant Dicer is able to function *in vitro* as a dsRNA-specific endonuclease, the *in vivo* experiments demonstrated that Dicer functions in association with other proteins as a component of multiprotein complexes. Dicer also plays a role in the assembly of RISC complexes. The Dicer-interacting proteins can be grouped into three broad classes: (i) dsRBD-containing proteins; (ii) Ago proteins; and (iii) other proteins, including helicase-related cofactors. Many of these proteins regulate Dicer-mediated processing of pre-miRNAs, often in a miRNA-specific way (for review, see Ref. [91]).

8.1. dsRBD proteins

Majority of Dicer proteins have been found to associate with dsRBD-containing cofactors, which modulate activity of the enzyme. The first to be described was the RNAi-deficient-4 (Rde-4) protein identified in a genetic screen in *C. elegans* for mutants deficient in RNAi and subsequently shown to interact with Dcr1 [92,93]. As it preferentially binds longer dsRNA and not siRNA, it is required for the initiation steps of RNAi [94]; Rde-4 is dispensable for miRNA processing and worm development [92,94]. In addition to discriminating between different substrates, Rde-4 has been demonstrated to activate Dcr1 [95]. In *Drosophila*, distinct dsRBD protein cofactors are required for the biogenesis and function of miRNAs and siRNAs (reviewed by Carthew and Sontheimer [96]). In miRNA biogenesis, depletion of the Dcr1-associated Loquacious (Loqs) protein results in the accumulation of pre-miRNAs, indicating it is essential for initial steps of processing by Dcr-1; the protein is not required for loading of mature miRNAs into Ago-1 [97–99]. Association of Loqs with Dcr-1 also activates the enzyme toward cleaving dsRNA [99]. For more details about Loqs, see Chapter 2. For siRNA biogenesis in *Drosophila*, Dcr-2 partners with R2D2, which is required for the initiation of silencing [100,101]. *In vitro* assays using siRNA duplexes demonstrated that R2D2 senses the thermodynamic asymmetry of the duplex by binding to the more stable end, thus positioning Dcr-2 at the opposite less stable end and helping in asymmetric siRNA loading into RISC [102]. Further *in vitro* studies using dsRNA substrates embedded with siRNAs of defined polarity, indicated that that the siRNA strand selection is independent of the initial dsRNA-processing polarity that generates siRNAs [103]. Consequently, the siRNA must first dissociate from Dicer before being repositioned in favorable polarity in the Dcr-2/R2D2 heterodimer. Similar siRNA repositioning may also occur after dsRNA cleavage by human Dicer [104] although evidence indicating that the dsRNA-processing polarity influences the siRNA strand selection in human cells has also been reported [105]. Together with Dcr-2, R2D2 is also involved in the sorting of small RNAs into different Ago-containing complexes [106,107]. *In vitro* analysis using purified recombinant proteins showed that Dcr-2/R2D2 heterodimer, but not Dcr-2 alone, directs the association of a small RNA duplex with Ago-2. Furthermore, the heterodimer was found to promote siRNA association with Ago-2 while at the same time disfavoring miRNA association [107]. The role of R2D2 in restricting substrate specificity of Dcr-2 was further strengthened when it was shown that, together with inorganic phosphate, the Dcr-2/R2D2 heterodimer inhibits pre-miRNA cleavage [10]. As discussed above, this work also demonstrated that the functional helicase/ATPase domain of Dcr-2 has evolved to process long dsRNAs in an ATP-dependent way and that, together with R2D2, this domain provides specificity for such substrates. Furthermore, analysis of R2D2 mutants demonstrated that siRNAs fail to load with Ago-2 but instead accumulated in the Ago-1 complex, thus confirming R2D2's central role in the sorting of small RNAs into different Ago complexes [106]. Finally, Loqs and R2D2 have been shown to also function sequentially and nonredundantly in the endo-siRNA pathway. Together with Dcr-2, Loqs is first required for processing of endogenous dsRNA and R2D2 then participates in loading of resulting siRNAs into RISC [108].

In mammals, both the HIV-1 TAR RNA-binding protein (TRBP) and the protein kinase R activator (PACT) have been shown to associate with

Dicer [47,109,110]. The two proteins show 42% identity and depletion of either of them results in defective RNA silencing [47,109–111]. Do TRBP or PACT play an analogous role to that of the *Drosophila* R2D2/Dcr-2 dimer in thermodynamically sensing the siRNA duplex and driving asymmetric strand loading into RISC? Although it has been reported that TRBP may function as a main siRNA asymmetry sensor [112], more recent experiments, performed with recombinant Dicer and TRBP, indicated that this role is performed by Dicer itself binding preferentially to the less stable siRNA end [104]. How the asymmetric siRNA binding to the Dicer–TRBP complex is contributing to the strand selection by RISC remains unclear since it has been demonstrated that Dicer is dispensable for asymmetric RISC loading both *in vitro* and *in vivo* in mammals [113]. TRBP has, however, been shown to enhance the stability and efficiency of both Dicer and RISC [109,110,114–117]. Analysis of dicing kinetics revealed that TRBP enhances the rate of cleavage of either pre-miRNAs or pre-siRNAs by approximately fivefold under multiple turnover conditions by increasing the stability of Dicer–substrate complexes [114]. Interestingly, TRBP phosphorylation was found to markedly enhance miRNA production by increasing stability of the Dicer–TRBP complex, without a change in cleavage kinetics [117].

TRBP contains two canonical dsRBDs at its N-terminus and a C-terminal domain that shares weak homology to a dsRBD but does not bind RNA. Stimulation of Dicer requires both N-terminal dsRBDs [114], whereas the C-terminal dsRBD of TRBP (and also of PACT) interacts with the helicase domain of Dicer [19,47,110,114,116]. In addition to interacting with Dicer, the C-terminal dsRBD also binds to the tumor suppressor Merlin as well as PACT and has been named the Merlin-Dicer-PACT liason (Medipal) domain [118]. Interestingly, in some colorectal cancers mutations that truncate TRBP have been identified that result in a defect in miRNAs biogenesis due to the destabilization of Dicer [119].

In the plant *A. thaliana*, the family of five closely related dsRNA-binding (DRB) proteins has been shown to interact with Dicers (DCLs). DRB1, also known as Hyponastic leaves 1 (HYL1), is one of the most intensively studied DCL-interacting proteins and is shown to be an important regulator of miRNA biogenesis [120,121]. *In vitro* processing assays, using recombinant DCL1, HYL1, as well as the associated Zn-finger protein SERRATE (SE), revealed that DCL1 alone is capable of substrate processing but the addition of HYL1 and SE proteins accelerates both the rate and accuracy of miRNA formation [120]. Roles for the other DRB proteins have also been described. Briefly, DRB2 also associates with DCL1 during miRNA biogenesis and may have both synergistic and antagonistic role with DRB1 in miRNA biogenesis [122]; DRB4 is required by DCL4 for the production of tasiRNAs [123] and its role can also be modulated by DRB2 [124]; and studies of various combinations of DRB2, DRB3, and DRB5 mutants suggested that DRB3 and DRB5 are together involved in mediating RNA silencing of target genes of the DRB2-controlled miRNAs [125].

8.2. Ago proteins

Ago proteins represent another group of well-characterized Dicer partners (see Chapter 4). Members of this superfamily generally contain four domains: an N-terminal (N), PAZ, middle (MID), and a C-terminal PIWI domain that is a "signature" of the Argonautes [126]. The proteins can be divided into three separate subgroups: the Ago clade containing members that are homologous to the *Arabidopsis* Ago-1 and associate with miRNAs and siRNAs, the Piwi clade that binds piwi-interacting RNAs (piRNAs), and a third clade that is specific to nematodes [127]. Many organisms express multiple members of the Ago superfamily proteins. For example, humans express eight (four of the Ago clade and four of the Piwi clade), *Drosophila* five, *C. elegans* 27, and *Arabidopsis* 10 proteins. Generally, individual family members perform specialized functions in RNA silencing. For example, in *Drosophila*, miRNAs and siRNAs are partitioned into effector complexes with different Ago proteins: miRNAs are loaded into an Ago-1-RISC, whereas siRNAs are incorporated into Ago-2-RISC [107,128]. Regardless of the type of the silencing pathway involving small RNAs, all are thought to require at least one or more Ago proteins that act as the central defining component of the various RISC and related effector complexes (see Chapter 4). Given that Dicer generates the products that enter into RISC or RITS, its relationship with Ago proteins has been intensively studied.

In mammals, four different Ago proteins are expressed with all four operating in the miRNA pathway, while only Ago-2 being able to function in RNAi by catalyzing the endonucleolytic cleavage of mRNA [129–131]. Human Dicer was shown to directly interact with Ago-2 via its RNase III domains binding to a subregion of the PIWI domain, termed the PIWI box [132]. In an early model of RISC formation also described above, Dicer together with Ago-2 and TRBP was proposed to first form a RISC-loading complex (RLC). Within RLC, Dicer would cleave the pre-miRNA and then, together with TRBP, would orientate the product before handing it over to Ago-2 [109,115,133]. Since this model was proposed, a key question has been how the handover of the cleaved product from Dicer to Ago-2 occurs, especially as small RNA duplex also has to be unwound to allow the sensing of asymmetry. Furthermore, RNAi can be triggered in Dicer knockout ES cells by transfecting miRNA duplexes, suggesting Dicer-independent loading of Ago-2 can occur [82]. Similar findings were made in flies when it was shown that neither Dcr-1 nor Dcr-2 is required for *in vitro* Ago-1-RISC assembly if miRNA duplexes are present [134].

Most recently, the relationship between human Dicer and Ago-2 has been intensively studied using *in vitro* RISC assembly assays with mammalian cell lysates programmed with pre-miRNAs instead of duplex miRNA [135]. Confirming earlier observations of a RLC-containing Dicer, Ago-2 and TRBP, small amounts of miRNA duplex were also found generated from the added pre-miRNA. Pulse-chase experiments confirmed that the duplex intermediate is converted to mature RISC, which contains Ago-2 and single-stranded miRNA; the conversion was dependent on the catalytic proficiency of Dicer. Interestingly, these assays also showed that certain pre-miRNAs directly bind to Ago-2, forming an miRNA deposit complex (miPDC) that ultimately loads into RISC [135]. The functionality of such an miPDC confirms previous observations that Ago-2 can bind and slice a subset of pre-miRNAs that may facilitate the removal of the passenger strand from mature RISC [136]. The Ago-2-dependent silencing independent of Dicer has been demonstrated for pre-miR-451 which is too short to be cleaved by Dicer but instead is directly bound by Ago-2 to enter RISC [137,138].

8.3. Other proteins interacting with dicer

Although Ago and dsRBD proteins are the best studied Dicer interactors, a growing list of other proteins have been identified which interact with the enzyme (for reviews, see Refs. [3,91,139,140] and references therein). For example, a proteomic screen in *C. elegans* revealed over 100 proteins associated with Dicer, 20 of which were demonstrated to reproducibly copurify with the enzyme [141]. Multiple classes of new proteins not previously known to be associated with Dicer were identified, including enhancers of RNAi (ERI) proteins, the PIR-1 group, and other proteins required for RNAi and development as well as proteins without a clearly defined function in RNA silencing. These findings revealed complexity of biochemical networks that Dicer is linked to [141]. Several of the proteins identified were previously known to interact with Dicer including the Dicer-related helicase (DRH) proteins whose roles are still poorly understood. These proteins contain a helicase domain which has a sequence and organization similar to the helicase domain of Dicer and RIG-I proteins. DRH-1 was shown to be required for RNAi and interact with Rde-4 and Dicer [93], whereas DRH-3 was found to be essential for RNAi in the germline [141].

9. FUTURE PERSPECTIVES

New studies continue to extend our understanding of Dicer function as well as our knowledge of the complexity of small RNA pathways (reviewed in Refs. [5,137,138]). For example, mammalian Dicer has been shown to have a role in processing of substrates others than dsRNA and pre-miRNAs. In addition to snoRNAs [78,79] and tRNAs [142], transcripts derived from Alu elements can also be processed by Dicer. Studies in mouse retina indicated that overaccumulation of *Alu* RNAs due to Dicer mutation is toxic to cells, causing cell death and age-related macular degeneration [143,144]. It will be important to reanalyze deep sequencing data to assess whether Dicer-dependent accumulation of Alu-derived small RNAs occurs in other cells or tissues. Various studies also describe that transcripts containing trinucleotide repeats, implicated in many neurological diseases, can form secondary structures that can be cleaved by Dicer [145–147]. In Huntington disease, Dicer was shown to process a subclass of triplet repeats, expansion of which is linked to the pathology of the disease [148]. It has also been suggested that Dicer might be involved in pre-rRNA processing as depletion of Dicer results in accumulation of precursors of 5.8S rRNA in cultured cells [149]. It will be important to obtain additional evidence in support of biological significance of the aforementioned processing reactions.

In *C. elegans*, proteolysis of Dicer has been suggested to inactivate the enzyme's RNase function, with the truncated enzyme gaining DNase activity and contributing to chromosomal breakup and apoptosis [150]. Proteolysis of Dicer in mammalian neurons has also been reported after neuronal stimulation possibly suggesting that Dicer may be involved in synaptic plasticity [38]. Furthermore, evidence exists that Dicer is alternatively spliced although the roles of the various Dicer isoforms remains to be determined [151–153]. It will be important to gain more knowledge about potential isoforms of Dicer and their physiological role. Understanding the role of Dicer in different cellular compartments will also be an important subject of future studies. Is Dicer involved in chromatin regulation also in mammalian cells? If so, what are its substrates in the nucleus? Is Dicer targeted to dendrites and axons of neuronal cells to participate in processing of pre-miRNAs reported as enriched in this compartment? [38,154–156].

Finally, as described here, much progress has been made in understanding the biochemistry of Dicer, with significant advances on how the enzyme's structure relates to its function. A key future task will be determining the X-ray structure of large canonical Dicers such as those expressed in different metazoa. Also, establishing 3D structures of Dicer-containing complexes, for example, heterodimers with the dsRBDs-containing partners, or RLC or RISC complexes at different stages of assembly, will represent an important future challenge.

ACKNOWLEDGMENTS

We thank Marc Buehler for his comments on the chapter. The Friedrich Miescher Institute for Biomedical Research is supported by the Novartis Research Foundation.

REFERENCES

[1] Fire A, et al. Potent and specific genetic interference by double-stranded RNA in *Caenorhabditis elegans*. Nature 1998;391(6669):806–11.
[2] Bernstein E, et al. Role for a bidentate ribonuclease in the initiation step of RNA interference. Nature 2001;409(6818):363–6.
[3] Jaskiewicz L, Filipowicz W. Role of Dicer in posttranscriptional RNA silencing. Curr Top Microbiol Immunol 2008;320:77–97.
[4] Jinek M, Doudna JA. A three-dimensional view of the molecular machinery of RNA interference. Nature 2009;457(7228):405–12.
[5] Kim VN, Han J, Siomi MC. Biogenesis of small RNAs in animals. Nat Rev Mol Cell Biol 2009;10(2):126–39.
[6] Drinnenberg IA, et al. RNAi in budding yeast. Science 2009;326(5952):544–50.
[7] MacRae IJ, et al. Structural basis for double-stranded RNA processing by Dicer. Science 2006;311(5758):195–8.
[8] Nicholson AW. Function, mechanism and regulation of bacterial ribonucleases. FEMS Microbiol Rev 1999;23(3):371–90.
[9] Blaszczyk J, et al. Crystallographic and modeling studies of RNase III suggest a mechanism for double-stranded RNA cleavage. Structure 2001;9(12):1225–36.
[10] Cenik ES, et al. Phosphate and R2D2 restrict the substrate specificity of Dicer-2, an ATP-driven ribonuclease. Mol Cell 2011;42(2):172–84.

[11] Lee YS, et al. Distinct roles for Drosophila Dicer-1 and Dicer-2 in the siRNA/miRNA silencing pathways. Cell 2004;117(1):69–81.

[12] Brodersen P, Voinnet O. The diversity of RNA silencing pathways in plants. Trends Genet 2006;22(5):268–80.

[13] Voinnet O. Origin, biogenesis, and activity of plant microRNAs. Cell 2009;136:669–87.

[14] Provost P, et al. Ribonuclease activity and RNA binding of recombinant human Dicer. EMBO J 2002;21(21):5864–74.

[15] Zhang H, et al. Human Dicer preferentially cleaves dsRNAs at their termini without a requirement for ATP. EMBO J 2002;21(21):5875–85.

[16] Zhang H, et al. Single processing center models for human Dicer and bacterial RNase III. Cell 2004;118(1):57–68.

[17] Guex N, Peitsch MC. SWISS-MODEL and the Swiss-PdbViewer: an environment for comparative protein modeling. Electrophoresis 1997;18(15):2714–23.

[18] Lau PW, et al. Structure of the human Dicer-TRBP complex by electron microscopy. Structure 2009;17(10):1326–32.

[19] Wang HW, et al. Structural insights into RNA processing by the human RISC-loading complex. Nat Struct Mol Biol 2009;16(11):1148–53.

[20] Sawh AN, Duchaine TF. Turning Dicer on its head. Nat Struct Mol Biol 2012;19 (4):365–6.

[21] Lau PW, et al. The molecular architecture of human Dicer. Nat Struct Mol Biol 2012;19(4):436–40.

[22] MacRae IJ, Zhou K, Doudna JA. Structural determinants of RNA recognition and cleavage by Dicer. Nat Struct Mol Biol 2007;14(10):934–40.

[23] Gan J, et al. Structural insight into the mechanism of double-stranded RNA processing by ribonuclease III. Cell 2006;124(2):355–66.

[24] Gan J, et al. A stepwise model for double-stranded RNA processing by ribonuclease III. Mol Microbiol 2008;67(1):143–54.

[25] Weinberg DE, et al. The inside-out mechanism of Dicers from budding yeasts. Cell 2011;146(2):262–76.

[26] Sun W, Pertzev A, Nicholson AW. Catalytic mechanism of Escherichia coli ribonuclease III: kinetic and inhibitor evidence for the involvement of two magnesium ions in RNA phosphodiester hydrolysis. Nucleic Acids Res 2005;33(3):807–15.

[27] Du Z, et al. Structural and biochemical insights into the dicing mechanism of mouse Dicer: a conserved lysine is critical for dsRNA cleavage. Proc Natl Acad Sci USA 2008;105(7):2391–6.

[28] Takeshita D, et al. Homodimeric structure and double-stranded RNA cleavage activity of the C-terminal RNase III domain of human dicer. J Mol Biol 2007;374 (1):106–20.

[29] Ye X, Paroo Z, Liu Q. Functional anatomy of the Drosophila microRNA-generating enzyme. J Biol Chem 2007;282(39):28373–8.

[30] Gurtan AM, et al. In vivo structure-function analysis of human Dicer reveals directional processing of precursor miRNAs. RNA 2012;18(6):1116–22.

[31] Ohishi K, Nakano T. A forward genetic screen to study mammalian RNA interference: essential role of RNase IIIa domain of Dicer1 in 3' strand cleavage of dsRNA in vivo. FEBS J 2012;279(5):832–43.

[32] Heravi-Moussavi A, et al. Recurrent somatic DICER1 mutations in nonepithelial ovarian cancers. N Engl J Med 2012;366(3):234–42.

[33] Park JE, et al. Dicer recognizes the 5' end of RNA for efficient and accurate processing. Nature 2011;475(7355):201–5.

[34] Ketting RF, et al. Dicer functions in RNA interference and in synthesis of small RNA involved in developmental timing in C. elegans. Genes Dev 2001;15(20):2654–9.

[35] Ma E, et al. Autoinhibition of human dicer by its internal helicase domain. J Mol Biol 2008;380(1):237–43.

[36] Ma E, et al. Coordinated activities of human dicer domains in regulatory RNA processing. J Mol Biol 2012;422(4):466–76.

[37] Soifer HS, et al. A role for the Dicer helicase domain in the processing of thermodynamically unstable hairpin RNAs. Nucleic Acids Res 2008;36 (20):6511–22.

[38] Lugli G, et al. Dicer and eIF2c are enriched at postsynaptic densities in adult mouse brain and are modified by neuronal activity in a calpain-dependent manner. J Neurochem 2005;94(4):896–905.

[39] Welker NC, et al. Dicer's helicase domain discriminates dsRNA termini to promote an altered reaction mode. Mol Cell 2011;41(5):589–99.

[40] Welker NC, et al. Dicer's helicase domain is required for accumulation of some, but not all *C. elegans* endogenous siRNAs. RNA 2010;16(5):893–903.

[41] Colmenares SU, et al. Coupling of double-stranded RNA synthesis and siRNA generation in fission yeast RNAi. Mol Cell 2007;27(3):449–61.

[42] Jiang F, et al. Structural basis of RNA recognition and activation by innate immune receptor RIG-I. Nature 2011;479(7373):423–7.

[43] Kowalinski E, et al. Structural basis for the activation of innate immune pattern-recognition receptor RIG-I by viral RNA. Cell 2011;147(2):423–35.

[44] Luo D, et al. Structural insights into RNA recognition by RIG-I. Cell 2011;147 (2):409–22.

[45] Dlakic M. DUF283 domain of Dicer proteins has a double-stranded RNA-binding fold. Bioinformatics 2006;22(22):2711–4.

[46] Qin H, et al. Structure of the *Arabidopsis thaliana* DCL4 DUF283 domain reveals a noncanonical double-stranded RNA-binding fold for protein-protein interaction. RNA 2010;16(3):474–81.

[47] Lee Y, et al. The role of PACT in the RNA silencing pathway. EMBO J 2006;25 (3):522–32.

[48] Clery A, Allain FH. From structure to function of RNA binding domains. In: Lorkovic Z, editor. RNA binding proteins. Landes Bioscience2012. p. 137–58.

[49] Ryter JM, Schultz SC. Molecular basis of double-stranded RNA-protein interactions: structure of a dsRNA-binding domain complexed with dsRNA. EMBO J 1998;17 (24):7505–13.

[50] Sun W, Jun E, Nicholson AW. Intrinsic double-stranded-RNA processing activity of *Escherichia coli* ribonuclease III lacking the dsRNA-binding domain. Biochemistry 2001;40(49):14976–84.

[51] Han J, et al. The Drosha-DGCR8 complex in primary microRNA processing. Genes Dev 2004;18(24):3016–27.

[52] Shi H, Tschudi C, Ullu E. An unusual Dicer-like1 protein fuels the RNA interference pathway in *Trypanosoma brucei*. RNA 2006;12(12):2063–72.

[53] Emmerth S, et al. Nuclear retention of fission yeast dicer is a prerequisite for RNAi-mediated heterochromatin assembly. Dev Cell 2010;18(1):102–13.

[54] Tsutsumi A, et al. Recognition of the pre-miRNA structure by *Drosophila* Dicer-1. Nat Struct Mol Biol 2011;18(10):1153–8.

[55] Volpe TA, et al. Regulation of heterochromatic silencing and histone H3 lysine-9 methylation by RNAi. Science 2002;297(5588):1833–7.

[56] Woolcock KJ, et al. Dicer associates with chromatin to repress genome activity in *Schizosaccharomyces pombe*. Nat Struct Mol Biol 2011;18(1):94–9.

[57] Buhler M. RNA turnover and chromatin-dependent gene silencing. Chromosoma 2009;118(2):141–51.

[58] Lejeune E, Allshire RC. Common ground: small RNA programming and chromatin modifications. Curr Opin Cell Biol 2011;23(3):258–65.

[59] Reyes-Turcu FE, Grewal SI. Different means, same end-heterochromatin formation by RNAi and RNAi-independent RNA processing factors in fission yeast. Curr Opin Genet Dev 2012;22(2):156–63.

[60] Bayne EH, et al. Stc1: a critical link between RNAi and chromatin modification required for heterochromatin integrity. Cell 2010;140(5):666–77.

[61] Gullerova M, Proudfoot NJ. Transcriptional interference and gene orientation in yeast: noncoding RNA connections. Cold Spring Harb Symp Quant Biol 2011;75:299–311.

[62] Gullerova M, Moazed D, Proudfoot NJ. Autoregulation of convergent RNAi genes in fission yeast. Genes Dev 2011;25(6):556–68.

[63] Barraud P, et al. An extended dsRBD with a novel zinc-binding motif mediates nuclear retention of fission yeast Dicer. EMBO J 2011;30(20):4223–35.

[64] Woolcock KJ, et al. RNAi keeps Atf1-bound stress response genes in check at nuclear pores. Genes Dev 2012;26(7):683–92.

[65] Grimaud C, et al. RNAi components are required for nuclear clustering of Polycomb group response elements. Cell 2006;124(5):957–71.

[66] Cernilogar FM, et al. Chromatin-associated RNA interference components contribute to transcriptional regulation in *Drosophila*. Nature 2011;480(7377):391–5.

[67] Matzke M, et al. RNA-mediated chromatin-based silencing in plants. Curr Opin Cell Biol 2009;21(3):367–76.

[68] Matzke MA, Birchler JA. RNAi-mediated pathways in the nucleus. Nat Rev Genet 2005;6(1):24–35.

[69] Zhang H, Zhu JK. Seeing the forest for the trees: a wide perspective on RNA-directed DNA methylation. Genes Dev 2012;26(16):1769–73.

[70] Park W, et al. CARPEL FACTORY, a Dicer homolog, and HEN1, a novel protein, act in microRNA metabolism in *Arabidopsis thaliana*. Curr Biol 2002;12(17):1484–95.

[71] Reinhart BJ, et al. MicroRNAs in plants. Genes Dev 2002;16(13):1616–26.

[72] Wassenegger M, et al. RNA-directed de novo methylation of genomic sequences in plants. Cell 1994;76(3):567–76.

[73] Mette MF, et al. Transcriptional silencing and promoter methylation triggered by double-stranded RNA. EMBO J 2000;19(19):5194–201.

[74] Hamilton A, et al. Two classes of short interfering RNA in RNA silencing. EMBO J 2002;21(17):4671–9.

[75] Xie Z, et al. Genetic and functional diversification of small RNA pathways in plants. PLoS Biol 2004;2(5):E104.

[76] Liu F, Bakht S, Dean C. Cotranscriptional role for *Arabidopsis* DICER-LIKE 4 in transcription termination. Science 2012;335(6076):1621–3.

[77] Ando Y, et al. Nuclear pore complex mediated nuclear localization of dicer protein in human cells. PLoS One 2011;6(8):e23385.

[78] Ender C, et al. A human snoRNA with microRNA-like functions. Mol Cell 2008;32 (4):519–28.

[79] Taft RJ, et al. Small RNAs derived from snoRNAs. RNA 2009;15(7):1233–40.

[80] Sinkkonen L, et al. Dicer is associated with ribosomal DNA chromatin in mammalian cells. PLoS One 2010;5(8):e12175.

[81] Benetti R, et al. A mammalian microRNA cluster controls DNA methylation and telomere recombination via Rbl2-dependent regulation of DNA methyltransferases. Nat Struct Mol Biol 2008;15(3):268–79.

[82] Kanellopoulou C, et al. Dicer-deficient mouse embryonic stem cells are defective in differentiation and centromeric silencing. Genes Dev 2005;19(4):489–501.

[83] Murchison EP, et al. Characterization of Dicer-deficient murine embryonic stem cells. Proc Natl Acad Sci USA 2005;102(34):12135–40.

[84] Kanellopoulou C, et al. X chromosome inactivation in the absence of Dicer. Proc Natl Acad Sci USA 2009;106(4):1122–7.

[85] Nesterova TB, et al. Dicer regulates Xist promoter methylation in ES cells indirectly through transcriptional control of Dnmt3a. Epigenetics Chromatin 2008;1(1):2.

[86] Ogawa Y, Sun BK, Lee JT. Intersection of the RNA interference and X-inactivation pathways. Science 2008;320(5881):1336–41.

[87] Fukagawa T, et al. Dicer is essential for formation of the heterochromatin structure in vertebrate cells. Nat Cell Biol 2004;6(8):784–91.

[88] Murchison EP, et al. Critical roles for Dicer in the female germline. Genes Dev 2007;21(6):682–93.

[89] Haussecker D, Proudfoot NJ. Dicer-dependent turnover of intergenic transcripts from the human beta-globin gene cluster. Mol Cell Biol 2005;25(21):9724–33.

[90] Sinkkonen L, et al. MicroRNAs control de novo DNA methylation through regulation of transcriptional repressors in mouse embryonic stem cells. Nat Struct Mol Biol 2008;15(3):259–67.

[91] Krol J, Loedige I, Filipowicz W. The widespread regulation of microRNA biogenesis, function and decay. Nat Rev Genet 2010;11(9):597–610.

[92] Tabara H, et al. The rde-1 gene, RNA interference, and transposon silencing in C. elegans. Cell 1999;99(2):123–32.

[93] Tabara H, et al. The dsRNA binding protein RDE-4 interacts with RDE-1, DCR-1, and a DExH-box helicase to direct RNAi in C. elegans. Cell 2002;109(7):861–71.

[94] Parker GS, Eckert DM, Bass BL. RDE-4 preferentially binds long dsRNA and its dimerization is necessary for cleavage of dsRNA to siRNA. RNA 2006;12(5):807–18.

[95] Parker GS, Maity TS, Bass BL. dsRNA binding properties of RDE-4 and TRBP reflect their distinct roles in RNAi. J Mol Biol 2008;384(4):967–79.

[96] Carthew RW, Sontheimer EJ. Origins and mechanisms of miRNAs and siRNAs. Cell 2009;136(4):642–55.

[97] Forstemann K, et al. Normal microRNA maturation and germ-line stem cell maintenance requires Loquacious, a double-stranded RNA-binding domain protein. PLoS Biol 2005;3(7):e236.

[98] Liu X, et al. Dicer-1, but not Loquacious, is critical for assembly of miRNA-induced silencing complexes. RNA 2007;13(12):2324–9.

[99] Saito K, et al. Processing of pre-microRNAs by the Dicer-1-Loquacious complex in Drosophila cells. PLoS Biol 2005;3(7):e235.

[100] Liu Q, et al. R2D2, a bridge between the initiation and effector steps of the Drosophila RNAi pathway. Science 2003;301(5641):1921–5.

[101] Pham JW, Sontheimer EJ. Molecular requirements for RNA-induced silencing complex assembly in the Drosophila RNA interference pathway. J Biol Chem 2005;280(47):39278–83.

[102] Tomari Y, et al. A protein sensor for siRNA asymmetry. Science 2004;306(5700):1377–80.

[103] Preall JB, et al. Short interfering RNA strand selection is independent of dsRNA processing polarity during RNAi in Drosophila. Curr Biol 2006;16(5):530–5.

[104] Noland CL, Ma E, Doudna JA. siRNA repositioning for guide strand selection by human Dicer complexes. Mol Cell 2011;43(1):110–21.

[105] Rose SD, et al. Functional polarity is introduced by Dicer processing of short substrate RNAs. Nucleic Acids Res 2005;33(13):4140–56.

[106] Okamura K, et al. R2D2 organizes small regulatory RNA pathways in Drosophila. Mol Cell Biol 2011;31(4):884–96.

[107] Tomari Y, Du T, Zamore PD. Sorting of *Drosophila* small silencing RNAs. Cell 2007;130(2):299–308.

[108] Marques JT, et al. Loqs and R2D2 act sequentially in the siRNA pathway in *Drosophila*. Nat Struct Mol Biol 2010;17(1):24–30.

[109] Chendrimada TP, et al. TRBP recruits the Dicer complex to Ago2 for microRNA processing and gene silencing. Nature 2005;436(7051):740–4.

[110] Haase AD, et al. TRBP, a regulator of cellular PKR and HIV-1 virus expression, interacts with Dicer and functions in RNA silencing. EMBO Rep 2005;6(10):961–7.

[111] MacRae IJ, et al. In vitro reconstitution of the human RISC-loading complex. Proc Natl Acad Sci USA 2008;105(2):512–7.

[112] Gredell JA, et al. Recognition of siRNA asymmetry by TAR RNA binding protein. Biochemistry 2010;49(14):3148–55.

[113] Betancur JG, Tomari Y. Dicer is dispensable for asymmetric RISC loading in mammals. RNA 2012;18(1):24–30.

[114] Chakravarthy S, et al. Substrate-specific kinetics of Dicer-catalyzed RNA processing. J Mol Biol 2010;404(3):392–402.

[115] Gregory RI, et al. Human RISC couples microRNA biogenesis and posttranscriptional gene silencing. Cell 2005;123(4):631–40.

[116] Kok KH, et al. Human TRBP and PACT directly interact with each other and associate with dicer to facilitate the production of small interfering RNA. J Biol Chem 2007;282(24):17649–57.

[117] Paroo Z, et al. Phosphorylation of the human microRNA-generating complex mediates MAPK/Erk signaling. Cell 2009;139(1):112–22.

[118] Laraki G, et al. Interactions between the double-stranded RNA-binding proteins TRBP and PACT define the Medipal domain that mediates protein-protein interactions. RNA Biol 2008;5(2):92–103.

[119] Melo SA, et al. A TARBP2 mutation in human cancer impairs microRNA processing and DICER1 function. Nat Genet 2009;41(3):365–70.

[120] Dong Z, Han MH, Fedoroff N. The RNA-binding proteins HYL1 and SE promote accurate in vitro processing of pri-miRNA by DCL1. Proc Natl Acad Sci USA 2008;105(29):9970–5.

[121] Hiraguri A, et al. Specific interactions between Dicer-like proteins and HYL1/DRB-family dsRNA-binding proteins in *Arabidopsis thaliana*. Plant Mol Biol 2005;57 (2):173–88.

[122] Eamens AL, et al. DRB2 is required for microRNA biogenesis in *Arabidopsis thaliana*. PLoS One 2012;7(4):e35933.

[123] Adenot X, et al. DRB4-dependent TAS3 trans-acting siRNAs control leaf morphology through AGO7. Curr Biol 2006;16(9):927–32.

[124] Pelissier T, et al. Double-stranded RNA binding proteins DRB2 and DRB4 have an antagonistic impact on polymerase IV-dependent siRNA levels in *Arabidopsis*. RNA 2011;17(8):1502–10.

[125] Eamens AL, Wook Kim K, Waterhouse PM. DRB2, DRB3 and DRB5 function in a non-canonical microRNA pathway in *Arabidopsis thaliana*. Plant Signal Behav 2012;7(10):1224–9.

[126] Tolia NH, Joshua-Tor L. Slicer and the argonautes. Nat Chem Biol 2007;3(1):36–43.

[127] Yigit E, et al. Analysis of the *C. elegans* Argonaute family reveals that distinct Argonautes act sequentially during RNAi. Cell 2006;127(4):747–57.

[128] Forstemann K, et al. *Drosophila* microRNAs are sorted into functionally distinct argonaute complexes after production by dicer-1. Cell 2007;130(2):287–97.

[129] Liu J, et al. Argonaute2 is the catalytic engine of mammalian RNAi. Science 2004;305 (5689):1437–41.

[130] Meister G, et al. Human Argonaute2 mediates RNA cleavage targeted by miRNAs and siRNAs. Mol Cell 2004;15(2):185–97.

[131] Rivas FV, et al. Purified Argonaute2 and an siRNA form recombinant human RISC. Nat Struct Mol Biol 2005;12(4):340–9.

[132] Tahbaz N, et al. Characterization of the interactions between mammalian PAZ PIWI domain proteins and Dicer. EMBO Rep 2004;5(2):189–94.

[133] Maniataki E, Mourelatos Z. A human, ATP-independent, RISC assembly machine fueled by pre-miRNA. Genes Dev 2005;19(24):2979–90.

[134] Kawamata T, Seitz H, Tomari Y. Structural determinants of miRNAs for RISC loading and slicer-independent unwinding. Nat Struct Mol Biol 2009;16(9):953–60.

[135] Liu X, et al. Precursor microRNA-programmed silencing complex assembly pathways in mammals. Mol Cell 2012;46(4):507–17.

[136] Diederichs S, Haber DA. Dual role for argonautes in microRNA processing and post-transcriptional regulation of microRNA expression. Cell 2007;131(6):1097–108.

[137] Cheloufi S, et al. A dicer-independent miRNA biogenesis pathway that requires Ago catalysis. Nature 2010;465(7298):584–9.

[138] Cifuentes D, et al. A novel miRNA processing pathway independent of Dicer requires Argonaute2 catalytic activity. Science 2010;328(5986):1694–8.

[139] Okamura K. Diversity of animal small RNA pathways and their biological utility. Wiley Interdiscip Rev RNA 2011;3(3):351–68.

[140] Winter J, et al. Many roads to maturity: microRNA biogenesis pathways and their regulation. Nat Cell Biol 2009;11(3):228–34.

[141] Duchaine TF, et al. Functional proteomics reveals the biochemical niche of *C. elegans* DCR-1 in multiple small-RNA-mediated pathways. Cell 2006;124(2):343–54.

[142] Cole C, et al. Filtering of deep sequencing data reveals the existence of abundant Dicer-dependent small RNAs derived from tRNAs. RNA 2009;15(12):2147–60.

[143] Kaneko H, et al. DICER1 deficit induces Alu RNA toxicity in age-related macular degeneration. Nature 2011;471(7338):325–30.

[144] Tarallo V, et al. DICER1 loss and Alu RNA induce age-related macular degeneration via the NLRP3 inflammasome and MyD88. Cell 2012;149(4):847–59.

[145] Handa V, Saha T, Usdin K. The fragile X syndrome repeats form RNA hairpins that do not activate the interferon-inducible protein kinase, PKR, but are cut by Dicer. Nucleic Acids Res 2003;31(21):6243–8.

[146] Krol J, et al. Ribonuclease dicer cleaves triplet repeat hairpins into shorter repeats that silence specific targets. Mol Cell 2007;25(4):575–86.

[147] Sobczak K, et al. RNA structure of trinucleotide repeats associated with human neurological diseases. Nucleic Acids Res 2003;31(19):5469–82.

[148] Banez-Coronel M, et al. A pathogenic mechanism in Huntington's disease involves small CAG-repeated RNAs with neurotoxic activity. PLoS Genet 2012;8(2): e1002481.

[149] Liang XH, Crooke ST. Depletion of key protein components of the RISC pathway impairs pre-ribosomal RNA processing. Nucleic Acids Res 2011;39(11):4875–89.

[150] Nakagawa A, et al. Caspase-dependent conversion of Dicer ribonuclease into a death-promoting deoxyribonuclease. Science 2010;328(5976):327–34.

[151] Irvin-Wilson CV, Chaudhuri G. Alternative initiation and splicing in dicer gene expression in human breast cells. Breast Cancer Res 2005;7(4):R563–R569.

[152] Potenza N, et al. A novel splice variant of the human dicer gene is expressed in neuroblastoma cells. FEBS Lett 2010;584(15):3452–7.

[153] Yan F, et al. Identification of novel splice variants of the *Arabidopsis* DCL2 gene. Plant Cell Rep 2009;28(2):241–6.

[154] Hengst U, et al. Functional and selective RNA interference in developing axons and growth cones. J Neurosci 2006;26(21):5727–32.

[155] Krichevsky AM, et al. A microRNA array reveals extensive regulation of microRNAs during brain development. RNA 2003;9(10):1274–81.

[156] Kye MJ, et al. Somatodendritic microRNAs identified by laser capture and multiplex RT-PCR. RNA 2007;13(8):1224–34.

Loquacious, a Dicer Partner Protein, Functions in Both the MicroRNA and siRNA Pathways

Ryuya Fukunaga*,†, Phillip D. Zamore*,†,1
*Howard Hughes Medical Institute, University of Massachusetts Medical School, Worcester, Massachusetts, USA
†Department of Biochemistry and Molecular Pharmacology, University of Massachusetts Medical School, Worcester, Massachusetts, USA
1Corresponding author: e-mail address: phillip.zamore@umassmed.edu

Contents

Abstract

In animals and plants, Dicer enzymes collaborate with double-stranded RNA-binding protein partners that change what substrates Dicer uses and where within the substrate they
dice. In the fruit fly *Drosophila melanogaster*, different Dicer enzymes produce microRNAs
(miRNAs) and small interfering RNAs (siRNAs): Dicer-1 produces miRNAs from pre-miRNAs,
while Dicer-2 generates siRNAs from long double-stranded RNA. Flies produce four Dicer
partner proteins—one encoded by the *r2d2* locus and three by *loquacious* (*loqs*). Alternative splicing of *loqs* generates Loqs-PA, Loqs-PB, and Loqs-PD. Loqs-PA and Loqs-PB
bind to Dicer-1 and assist miRNA biogenesis by increasing substrate affinity or enzyme
turnover. Loqs-PB also alters the size of some miRNAs, generating isoforms with seed
sequences and target specificities distinct from those that would be made by Dicer-1 alone
or Dicer-1 bound to Loqs-PA. Loqs-PD binds Dicer-2, promoting the production of both
endogenous and exogenous siRNAs by increasing substrate affinity. In this chapter, we will
discuss the diverse functions *loqs* plays in small silencing RNA pathways.

1. INTRODUCTION

Small silencing RNAs direct Argonaute proteins to repress partially or
fully complementary mRNA targets. Of the three classes of small silencing
RNAs, two—microRNAs (miRNAs) and small interfering RNAs
(siRNAs)—are generated by the RNase III enzyme Dicer, while Piwi-
interacting RNAs are generated in a Dicer-independent pathway. In animals
and plants, Dicer enzymes collaborate with double-stranded RNA-binding
protein partners that change what substrates Dicer uses and where within the
substrate they dice. These partner proteins comprise two or three tandem
dsRNA-binding domains (dsRBDs), a broadly conserved structural domain
that recognizes the shape but not the sequence of an A-form RNA helix and
that can also serve to mediate protein–protein interactions.

Associations between Dicer enzymes and dsRBD partner proteins are a general theme in animals and plants. A forward genetic screen for genes required for RNAi in the roundworm *Caenorhabditis elegans* identified the first Dicer partner protein, RNAi *de*fective-4 (RDE-4). RDE-4, which has two dsRBDs, binds to Dicer and is required for the production of siRNA [1,2]. In mammals, the paralogous proteins TAR RNA-binding protein (TRBP) and protein activator of PKR (PACT), each of which has three dsRBDs, bind to Dicer and promote the production of miRNA and siRNA [3–7].

The pairing of Dicer enzymes with dsRBD partner proteins is particularly specialized in plants. For example, each of the four *Arabidopsis thaliana* Dicers collaborates with its own individual dsRBD partner protein [8–10]. Dicer-like protein 1 (DCL1) associates with the dsRBD protein HYL1 (DRB1), which is required for efficient and precise production of miRNA [11–13]. HYL1 is also required for guide strand selective loading into an Argonaute silencing complex [8]. Plant DCL1 cleaves primary miRNA into pre-miRNA and then pre-miRNA into mature miRNA, whereas animal Dicer performs only the latter reaction and another RNase III enzyme, Drosha, performs the former reaction (Fig. 2.1A). Like HYL1, DRB2, DRB3, and DRB5 are implicated in the miRNA pathway [14,15]. Direct interactions between Dicer-like proteins 2 and 3 (DCL2 and DCL3) and DRB2, DRB3, and DRB5 have not been shown, but are likely. Dicer-like protein 4 (DCL4) binds to DRB4, which is crucial for the production of siRNA.

The first *Drosophila* Dicer partner protein to be identified was R2D2, named for its two dsRBDs and its association with Dicer-2. R2D2 forms a stable 1:1 heterodimer with Dicer-2 and functions in the siRNA pathway [16]. R2D2 has both positive and negative functions in small RNA biogenesis: it promotes the transfer of the siRNA guide strand from Dicer-2 to Argonaute2 (Ago2) [16–19], prevents endo-siRNAs sorting into Argonaute1 (Ago1) [20], and prevents Dicer-2 from processing miRNA precursors [21].

Drosophila genes are traditionally named to describe their loss-of-function phenotype. Loss of the *r2d2* paralog *loquacious* disrupts miRNA and siRNA production and causes a loss of silencing. *loqs* mutant flies do not efficiently produce miRNAs or siRNAs, causing a loss of silencing of the genes they regulate. Hence, the name "*loquacious*," meaning quite talkative [22–28]. Alternative splicing of the *loqs* primary transcript generates three distinct Dicer partner proteins (Loqs-PA, Loqs-PB, and Loqs-PD). Both Loqs-PA and Loqs-PB contain three dsRBDs, bind to Dicer-1, and enhance production of miRNA. Loqs-PB further acts to alter the nucleotide at

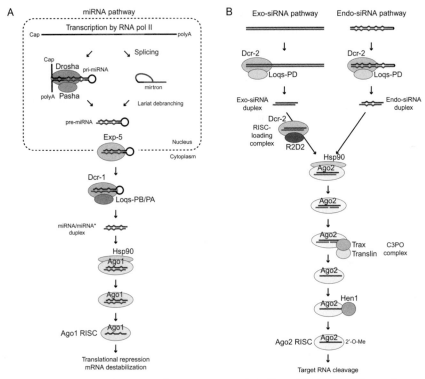

Figure 2.1 miRNA and siRNA pathways in *Drosophila*. The miRNA and siRNA pathways differ in their substrates, mechanisms of biogenesis, and modes of target silencing. (A) miRNAs are transcribed by RNA polymerase II (RNA pol II) as primary miRNAs (pri-miRNAs) bearing a 7-methyl guanosine cap and poly(A) tail. Drosha, assisted by its dsRBD-binding partner Pasha, cleaves canonical pri-miRNAs into precursor miRNAs (pre-miRNAs). A few pre-miRNAs (mirtrons) correspond to introns and are excised from their pri-miRNAs by the pre-mRNA splicing machinery. Exportin-5 (Exp-5) transfers pre-miRNAs from the nucleus to the cytoplasm, where they are cleaved by the Dicer-1 (Dcr-1): Loquacious-PA (Loqs-PA) or Dicer-1:Loqs-PB complex into miRNA/miRNA* duplexes. These duplexes are loaded into Argonaute1 (Ago1) with the aid of heat-shock protein 90 (Hsp90) and Hsc70 (Ago1 pre-RISC). The complex becomes mature Ago1-RISC after the miRNA* is released and can now regulate seed-matched targets by translational repression and mRNA destabilization. (B) The Dicer-2 (Dcr-2):Loqs-PD complex to generate exo-siRNA and endo-siRNA duplexes from long dsRNA and long RNA hairpins. The Dicer-2: R2D2 heterodimer (also called the RISC-loading complex) then loads the exo-siRNA duplex into Argonaute2 (Ago2) with help from Hsc70/Hsp90, generating pre-RISC. Ago2 cleaves the siRNA passenger strand, allowing the cleaved fragments to be removed by the Translin–Trax complex (C3PO complex). The final step in RISC maturation is the addition by the enzyme Hen1 of a 2′-*O*-methyl group to the 3′ end of the siRNA guide strand. The mature Ago2-RISC comprising Ago2 and a guide siRNA can then cleave extensively complementary RNA targets, such as viral mRNAs. (See color plate section in the back of the book.)

which Dicer-1 cleaves in several pre-miRNAs, and is thus required to produce the correct isoform of some miRNAs [28]. Loqs-PD has two dsRBDs, binds to Dicer-2, and promotes production of siRNA [25–29].

In this chapter, we will discuss the diverse functions of *loqs* in small silencing RNA pathways in *Drosophila*.

2. miRNA AND siRNA PATHWAYS IN THE FRUIT FLY *DROSOPHILA MELANOGASTER*

2.1. miRNA pathway

Distinct *Drosophila* Dicer enzymes generate 21-nt siRNAs and $\sim$21–24-nt miRNAs (Fig. 2.1). miRNAs are first transcribed as long primary transcripts (pri-miRNAs) by RNA polymerase II (Fig. 2.1A). The pri-miRNA is cleaved in the nucleus by the RNase III enzyme Drosha, aided by its dsRBD partner protein Pasha, into an $\sim$60–70-nt long hairpin RNA—a pre-miRNA—that bears the hallmarks common to all RNase III products: 2-nt 3'-overhang ends [30–34] (Figs. 2.1A and 2.2). A few pre-miRNAs, called mirtrons, are produced by splicing and debranching instead of Drosha cleavage [35,36]. The Exportin-5/Ran-GTP complex transports each pre-miRNA from the nucleus to the cytoplasm [37–39]. In cytoplasm, fly Dicer-1, aided by Loqs-PA or Loqs-PB, cleaves the pre-miRNA into $\sim$22-nt long duplex RNA comprising $\sim$20 bp with 2-nt 3'-overhang ends. The two strands of this duplex correspond to the mature miRNA and its miRNA⋆, a partially complementary small RNA derived from the opposite arm of the pre-miRNA stem [40].

The intact miRNA–miRNA⋆ duplex is loaded into Ago1. Biogenesis and Argonaute loading of miRNA/siRNA are not coupled and instead are independent processes [18,19]. Mismatches in the central region in the miRNA–miRNA⋆ duplex and an initial uridine nucleotide of the miRNA strand promote loading in Ago1. In contrast, base-pairing in the central region, and an initial cytidine nucleotide enhance Ago2 loading of siRNA duplex and miRNA–miRNA⋆ duplex in such orientation that the miRNA⋆ strand, instead of the miRNA strand, is maintained at a later step [18,19,41–43]. In general, Argonaute loading requires protein chaperones (heat-shock proteins) such as Hsc70/Hsp90 [44–46], presumably to accelerate the conversion from a closed conformation to a more open structure receptive to binding the duplex. The Ago1 complex only becomes functional after the miRNA⋆ strand dissociates. Specific mismatches between the miRNA and miRNA⋆, particularly in the miRNA seed

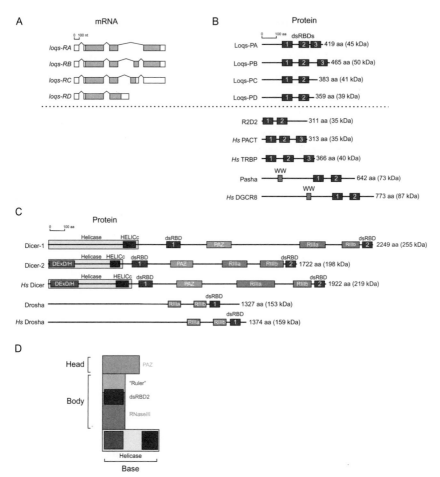

Figure 2.2 Loquacious isoforms. (A) The isoforms of *Drosophila loqs* mRNA. (B) Domain structures of the isoforms of *Drosophila* Loqs protein. Other dsRBD partners for the RNase III enzymes Drosha and Dicer are shown below. The tryptophan–tryptophan (WW) motif in Pasha and DGCR8 binds to a proline-rich motif in Drosha. (C) Domain structures of *Drosophila* and human Dicer and Drosha. DEAD, DEAD/DEAH box helicase domain; HELICc, helicase conserved C-terminal domain; PAZ, PAZ domain; RIIIa and RIIIb, Ribonuclease III domain; dsRBD, dsRNA-binding domain. (D) Cartoon model of the L-shaped structure of human Dicer. (See color plate section in the back of the book.)

sequence, promote release of the miRNA★ strand, generating an RNA-induced silencing complex (RISC) [18,47]. Ago1 RISC binds target mRNAs through base complementarity and suppresses protein translation and triggers mRNA degradation [48–50]. Repression by miRNAs plays critical roles in cell physiology, differentiation, proliferation, and disease.

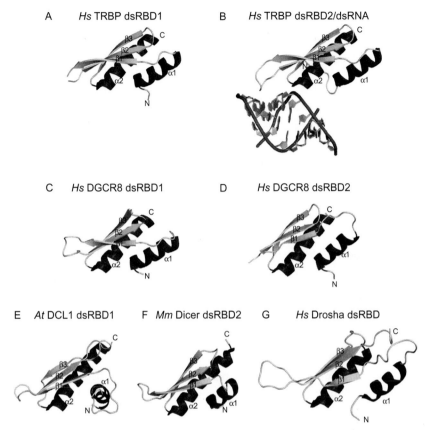

Figure 2.3 Structures of the dsRBDs from Dicer and Dicer/Drosha-binding partner proteins. α-helices are shown in blue and β-sheets in yellow. RNA is red or pink. (A) Crystal structure of human TRBP dsRBD1 (PDB ID: 3LLH) [87]. (B) Crystal structure of human TRBP dsRBD2 in complex with a 10 bp dsRNA (PDB ID: 3ADL) [88]. (C) Crystal structure of human DGCR8 dsRBD1 (PDB ID: 2YT4) [89]. (D) Crystal structure of human DGCR8 dsRBD2 (PDB ID: 2YT4) [89]. (E) NMR structure of *Arabidopsis* DCL4 dsRBD1 (PDB ID: 2KOU) [90]. (F) Crystal structure of mouse Dicer dsRBD2 (PDB ID: 3C4B) [91]. (G) NMR structure of human Drosha dsRBD (PDB ID: 2KHX) [92]. (See color plate section in the back of the book.)

2.2. siRNA pathway

Drosophila Dicer-2 produces siRNAs from both exogenous and endogenous dsRNA (Fig. 2.1B). Exogenous siRNAs (exo-siRNAs) in flies are thought to derive from long dsRNA molecules generated during viral replication [51]. exo-siRNAs can also be produced by introducing long dsRNA experimentally or by engineering transgenes that express long, fully

complementary hairpin transcripts [52–54]. Endogenous siRNAs (endo-
siRNAs) are derived from self-complementary hairpin transcripts
(esiRNAs), convergent mRNAs (*cis*-natural antisense transcripts (*cis*-NAT)
endo-siRNAs), or dsRNA from mobile elements [55–61].

Dicer-2 associates with the dsRNA-binding partners Loquacious-PD
(Loqs-PD) and R2D2, both of which contain two dsRBDs (Figs. 2.1B
and 2.2B) [16,17,25–27]. R2D2 is a paralog of Loqs (25% identity and
37% similarity in the regions of the two dsRBDs). The relative
thermodynamic stabilities of each end region of an siRNA or miRNA/
miRNA* determines which strand becomes the guide for RISC: the
strand with the less stable $5'$ is favored [62–64]. The Dicer-2:R2D2
heterodimer senses the thermodynamic asymmetry of exo-siRNA
duplexes and, together with Hsc70/Hsp90, loads them into Ago2 so that
the favored strand binds tightly to the Argonaute protein
[16–18,44–46,65,66]. The other strand of the siRNA, the passenger strand,
is then cleaved by Ago2, triggering its dissociation and destruction. The
Translin–Trax complex (C3PO complex) may facilitate release of the
cleaved passenger strand fragments [67–69]. Finally, the Ago2-bound guide
strand is $2'$-O-methylated at it $3'$ end by the methyltransferase Hen1 [70].
The resulting mature Ago2-RISC silences highly complementary target
RNAs by cleaving them at the phosphodiester bond flanked by guide
nucleotides 10 and 11, a process typically called RNAi [71,72].

3. *DROSOPHILA* DICER-1 AND DICER-2: SIMILAR BUT DISTINCT DOMAIN STRUCTURES

3.1. General domain organization of dicers

Drosophila Dicer-1, Dicer-2, and human Dicer have similar domain organization. Each has an amino-terminal helicase domain (composed of a Helicase C
domain with or without a functional DExD/H domain), a central atypical
dsRBD, a PAZ domain, two RNase III domains, and a carboxy-terminal
dsRBD (Fig. 2.2C). The two RNase III domains (RNase IIIa and RNase
IIIb) form an intramolecular heterodimer [73,74]. Cleavage by RNase III
domains results in a 2-nt $3'$-overhang, a $5'$-terminal phosphate, and a $3'$-
hydroxyl in the product. PAZ domains, which are found in both Dicer
and Argonaute proteins, recognize the characteristic 2-nt $3'$-overhang and
$5'$-terminal phosphate left by Drosha and Dicer cleavage [74–83]. The
distance between the PAZ domain and the active sites of the RNase III
domains serves as a molecular ruler that establishes the length of the small

RNA product [73,74,79,84–86]. NMR analysis revealed that the central atypical dsRBD, formerly a domain of unknown function (DUF283), adopts an α-β-β-β-α fold similar to that of a canonical dsRBD (Fig. 2.3) [90].

In reconstructions of negative stain electron microscopy, the entire Dicer protein forms an L-shaped structure [7,93,94]. The two RNase III domains and the C-terminal dsRBD are located in the body, and the PAZ domain is positioned at the head of the longer axis of the L. The helicase domain forms a clamp-shaped structure at base.

3.2. Domain organization of Dicer-1

In Dicer-1, the RNase IIIa and IIIb domains cleave the 3′ arm and 5′ arms of pre-miRNA [95], while the PAZ domain recognizes the 2-nt 3′-overhang and 5′-terminal phosphate of the pre-miRNA [83]. Dicer-1 has a helicase C domain, but lacks a functional DExD/H domain (Fig. 2.2C). The DExD/H domain is found in a wide range of "helicase" domains that couple ATP hydrolysis to binding, unwinding, or translocation along RNA or DNA molecules; such proteins can also rearrange RNA:protein, DNA:protein, or protein:protein interactions or act as RNA chaperones [96–102]. Consistent with the absence of a functional DExD/H domain, Dicer-1 does not require ATP to bind or cleave pre-miRNA [103]. The Dicer-1 helicase domain recognizes the pre-miRNA loop and confirms substrate authenticity by measuring the distance between the 3′-overhanging end and the loop and checking the loop size [103].

3.3. Domain organization of Dicer-2

In contrast to the helicase domain of Dicer-1, that of Dicer-2 has a functional DExD/H domain (Fig. 2.2C). In fact, long dsRNA stimulates Dicer-2 to hydrolyze ATP to ADP [21]. ATP binding and hydrolysis are required for Dicer-2 to processively cleave long dsRNA, but not to cleave short dsRNAs such as pre-miRNAs. Dicer-2 hydrolyzes $\sim$23 ATP molecules for each 21-nt siRNA made from long dsRNA, suggesting that Dicer-2 consumes approximately one ATP molecule to traverse each base pair along the dsRNA helix. Consequently, Dicer-2 is processive: once it binds to an end of dsRNA and starts to cleave a molecule of long dsRNA, it does not dissociate until it dices successively to the other end [21,104].

Pre-miRNA cleavage occurs only once (two strands cleaved) per pre-miRNA molecule and thus does not involve translocation of Dicer-1. The lack of a functional DExD/H domain in Dicer-1 may help Dicer-1 to cleave only its authentic substrate, pre-miRNA, and not long dsRNA. Dicer-2 on its own efficiently cleaves pre-miRNA, but this undesirable activity is thought to be suppressed *in vivo* by the physiological concentration of inorganic phosphate and by its dsRBD partner protein, R2D2 [21]. Dicer-2 binds to R2D2 through its helicase domain [29]. Thus, the distinct Dicer-1 and Dicer-2 helicase domains confer their distinct substrate specificities.

4. CONSERVED PARTNERSHIPS BETWEEN DICER AND DICER-BINDING PARTNER dsRBD PROTEINS

4.1. dsRBD partner proteins in animals and plants

Although the RNase III enzymes involved in small silencing RNA pathways contain their own dsRBDs (one in Drosha and two in Dicers; Fig. 2.2C), they still associate with dsRBD protein partners. In flies, Drosha collaborates with Pasha, Dicer-1 associates with Loqs-PA and Loqs-PB, and Dicer-2 binds to Loqs-PD and R2D2. In mammals, small silencing RNA production is likewise catalyzed by RNase III enzymes containing their own dsRBDs, aided by dsRBD partner proteins (Fig. 2.2B and C). Mammalian Drosha collaborates with DiGeorge syndrome critical region gene 8 (DGCR8), which, like its homolog Pasha, contains two dsRBDs [32]. Mammalian Dicer, which produces both miRNAs and siRNAs, associates with two alternative partner proteins, TRBP and PACT, each of which contains three dsRBDs (Fig. 2.2B) [3,4,32,105]. The amino acid sequences of these two dsRBD proteins suggest that they are orthologous to Loqs-PB. The amino acid sequence of TRBP has 24% identity and 33% similarity, while PACT has 28% identity and 34% similarity, to Loqs-PB.

Partnerships between Dicer enzymes and dsRBD protein partners appear to be a general theme in animals and plants. In contrast, no dsRBD partner proteins have been reported for Dicer in such unicellular eukaryotes as *Schizosaccharomyces pombe* (fission yeast), *Saccharomyces castellii* (budding yeast), *Giardia lamblia* (protozoan parasite), or *Tetrahymena thermophila* (ciliate protoza). Notably, the Dicer proteins in these organisms lack a helicase domain, suggesting that the Dicer helicase domain and Dicer dsRBD partner proteins may have coevolved. While bacteria and Archaea contain Argonaute homologs, no Dicer-like enzymes or Dicer partner proteins have been found in these kingdoms. Thus, dsRBD partner proteins may be an innovation found only in animals and plants.

4.2. Structure of dsRBDs

The dsRBD is about 70-amino acid (aa) long and adopts an α-β-β-β-α structure (Fig. 2.3). The two α helices pack against one face of the three-stranded antiparallel β sheet. The structures of dsRBD1 and dsRBD2 (in complex with 10 bp dsRNA) from human TRBP, dsRBD1 and dsRBD2 from human DGCR8, dsRBD1 from *Arabidopsis* DCL4, dsRBD2 from mouse Dicer, dsRBD from human Drosha are all similar, sharing the canonical α-β-β-β-α fold [87–92].

The dsRBD has a conserved positive electrostatic potential charge around the N-terminal part of $\alpha 2$, the region that recognizes the major groove of dsRNA (Fig. 2.3B). In addition, the loop between $\beta 1$ and $\beta 2$ has a positively charged patch that recognizes the dsRNA minor groove. The dsRBD generally recognizes dsRNA structure but not sequence. Proteins often contain tandem dsRBDs, as in Loqs, R2D2, TRBP, PACT, Pasha, and DGCR8 (Fig. 2.2). In human TRBP, dsRBD1 and dsRBD2 have been suggested to simultaneously bind to one dsRNA molecule, increasing its affinity for dsRNA [87].

5. *LOQS* ENCODES FOUR ALTERNATIVELY SPLICED ISOFORMS

The *loqs* (CG6866) gene in *Drosophila* lies on the left arm of Chromosome 2 at polytene band 34B9. Alternative splicing of *loqs* produces four mRNA isoforms, *loqs-RA*, *loqs-RB*, *loqs-RC*, and *loqs-RD* (Fig. 2.2A), each encoding a distinct Loqs protein isoform: Loqs-PA, Loqs-PB, Loqs-PC, and Loqs-PD (Fig. 2.2B). The first three isoforms were discovered in 2005 [22–24], while Loqs-PD was found later [25,26]. The longest Loqs protein isoform, Loqs-PB (465 aa, 50 kDa), comprises three dsRBDs. The first and second dsRBDs have canonical dsRBD sequences and are most closely related to dsRBD1 and dsRBD2 of R2D2. The third dsRBD (dsRBD3) of Loqs-PB deviates from the canonical dsRBD sequence and is required for Loqs-PB to bind to Dicer-1 [22,95]. *loqs-RA* lacks one exon present in *loqs-RB*, and Loqs-PA (419 aa, 45 kDa), which also binds to Dicer-1, lacks 46 aa found within the dsRBD3 of Loqs-PB. *loqs-RC* has a 5' extended fourth exon that changes its reading frame and truncates the Loqs-PC protein (383 aa, 41 kDa). Loqs-PC lacks the entire dsRBD3; instead it has a unique 46 aa carboxy-terminal sequence. The *loqs-RC* pre-mRNA contains an intron in its 3' untranslated region

(UTR), and thus the mature *loqs-RC* mRNA has an exon–exon junction in its 3' UTR after the stop codon. Such mRNAs are typically recognized as aberrant and targeted for destruction by the nonsense-mediated decay pathway [106]. In fact, Loqs-PC has not been detected in any cell or tissue from fly. (The S2 cell protein, originally thought to be Loqs-PC, we now know is Loqs-PD because RNAi depletion of *loqs-RD*, but not *loqs-RC*, eliminates it [22,26].) *loqs-RD* has a 3' extended third exon that contains a poly(A) signal sequence different from those used in *loqs-RA*, *-RB*, and *-RC*. Loqs-PD (359 aa, 39 kDa) has a unique 22 aa carboxy-terminal sequence in place of dsRBD3, allowing it to bind to Dicer-2 rather than Dicer-1 [25,26]. Ectopically expressed Loqs-PA, Loqs-PB, and Loqs-PD distribute evenly in cytosol in S2 cells, while ectopically expressed Loqs-PC aggregates as cytoplasmic foci [23,27]. Similarly, transgenic Loqs-PA, Loqs-PB, and Loqs-PD, but not Loqs-PC, can be expressed in the soma and germline of adult flies, and recombinant Loqs-PA, Loqs-PB, and Loqs-PD can be expressed as soluble proteins in *Escherichia coli*, while recombinant Loqs-PC produces aggregates [28] (Ryuya Fukunaga and Phillip Zamore, unpublished results). Collectively, these results suggest that *loqs-RC* does not encode a functional protein.

Isoform expression of *loqs* differs among fly tissues and among developmental stages. Expression levels of *loqs-RA* are lower at earlier stages of development and increase with developmental time [28]. In adult flies, both *loqs-RA* and *loqs-RB* are expressed in the carcasses of males and females after removal of the gonads. *loqs-RA* predominates in testes, whereas *loqs-RB* is the most abundant species in ovaries [22,27]. *loqs-RD* is the second most abundant isoform in both testes and ovaries. *loqs-RA*, *loqs-RB*, and *loqs-RD* are all expressed in S2 cells with *loqs-RA* being the least abundant. Expression of Loqs protein isoforms is consistent with mRNA isoform expression. Thus, the alternative splicing of the *loqs* pre-mRNA is regulated. How alternative splicing of *loqs* is regulated to produce *loqs* isoforms awaits future study.

6. LOQS-PA AND LOQS-PB BIND TO DICER-1, WHILE LOQS-PD AND R2D2 BIND TO DICER-2

6.1. Loqs-PA and Loqs-PB bind to Dicer-1

Immunoprecipitation experiments demonstrated that Dicer-1 binds to Loqs-PA and Loqs-PB, but not Loqs-PD [22–24, 95]. Endogenous Loqs-PA and Loqs-PB also copurify with Dicer-1 by cation exchange, anion exchange, hydroxyapatite, and gel filtration chromatography [22,24]. The stoichiometry of the Dicer-1:Loqs-PB complex formed using purified recombinant proteins is 1:1 [28]. Loqs-PA and Loqs-PB do not detectably bind to Dicer-2.

Experiments using a series of truncated proteins show interaction between the helicase domain of Dicer-1 and dsRBD3 of Loqs-PA and Loqs-PB [95]. Competition experiments between the dsRBD3s of Loqs-PA and the Loqs-PB for binding to Dicer-1 suggest that Loqs-PB binds Dicer-1 more tightly. When both His-tagged Loqs-PA and His-tagged Loqs-PB were coexpressed in S2 cells at similar levels, together with FLAG-tagged Dicer-1, more Loqs-PB than Loqs-PA copurified with Dicer-1 anti-FLAG immunoprecipitates [95], again suggesting that Dicer-1 has a higher affinity for Loqs-PB than Loqs-PA. Likely, the 46 amino acids missing from dsRBD3 of Loqs-PA but present in Loqs-PB explain the higher affinity of Loqs-PB for Dicer-1. The critical role of dsRBD3 in binding Dicer-1 also explains why Dicer-1 does not bind Loqs-PD, which lacks dsRBD3.

6.2. Loqs-PD and R2D2 bind to Dicer-2

Binding of Loqs-PD to Dicer-2, but not to Dicer-1, was shown using immunoprecipitation from S2 cells [25,27,29]. Just as Loqs-PA and Loqs-PB bind the helicase domain of Dicer-1, Loqs-PD binds the helicase domain of Dicer-2. The unique carboxy-terminal 22 aa of Loqs-PD plays a crucial role in this interaction. One study suggests that the 22 aa directly interacts with the Dicer-2 helicase domain [29], while another suggests that they help the extended dsRBD2 of Loqs-PD fold [27]. The special role for the unique 22 aa of the Loqs-PD dsRBD2 explains why neither Loqs-PA nor Loqs-PB binds to Dicer-2. Immunoprecipitates using anti-Myc antibody in S2 cells expressing Myc-tagged Loqs-PD contained R2D2 as well as Dicer-2, suggesting that the three proteins may form a ternary complex [27].

The helicase domain of Dicer-2 also binds to R2D2 [29]. Both dsRBDs of R2D2 are important for this interaction [65]. A second study, however, suggests that the carboxy-terminal region of the R2D2, not its dsRBDs, interacts with Dicer-2 [107]. *In vivo*, R2D2 is unstable in the absence of Dicer-2. Depletion of Dicer-2 from S2 cells using RNAi causes a large decrease in R2D2 protein abundance without changing the level of *r2d2* mRNA [16]. Similarly, *Dicer-2* null mutant flies have much less R2D2 protein than control flies [65]. In contrast, binding of Loqs-PA or Loqs-PB to Dicer-1 or binding of Loqs-PD to Dicer-2 does not seem to affect the stability of these proteins. In S2 cells depleted of Dicer-1 or Dicer-2 proteins by RNAi, Loqs protein levels are similar to control levels [22–24]. Reciprocally, Loqs protein depletion affects neither Dicer-1 nor Dicer-2 abundance. *In vivo*, however, in $loqs^{f00791}$ hypomorphic mutant fly ovaries and in the $loqs^{KO}$ null mutant embryo, Dicer-1 protein levels are reduced compared to controls [22,108], although Dicer-2 levels in the $loqs^{f00791}$ mutant flies are unchanged (Fukunaga and Zamore, unpublished results).

7. PARTNER PROTEIN MUTANT PHENOTYPES IN FLIES AND MAMMALS

7.1. *loqs* mutant phenotypes

A hypomorphic mutant allele of *loqs*, $loqs^{f00791}$, was created in a large-scale piggyBac transposon mutagenesis screen of *Drosophila*. The f00791 piggyBac lies in the *loqs* promoter, before the transcriptional start site. Compared with control flies, the hypomorphic mutant allele reduces the amount of *loqs* mRNA from 3- to 40-fold among different tissues in males and females [22]. The $loqs^{f00791}$ homozygous mutant flies are viable but females are sterile and males show reduced fertility. Precise excision of the f00791 piggyBac transposon fully rescued these phenotypes, confirming that the observed phenotypes are caused by the piggyBac insertion. $loqs^{f00791}$ females lay no eggs, and their ovaries are much smaller than normal, likely because they fail to maintain germline stem cells [22,24]. In the $loqs^{f00791}$ flies, mature miRNAs decrease and pre-miRNAs accumulate for some miRNAs, including miR-7 and miR-277, and esiRNAs decreased, compared with control flies [22–24].

A knockout, null mutant allele of *loqs*, $loqs^{KO}$, was generated by removing the entire *loqs* open reading frame by "ends-out" targeted homologous recombination [109]. $loqs^{KO}$ homozygous mutant embryos hatch and develop, and the mutant larvae pupariate at rates similar to wild type. However, viability dramatically drops at the pupal-to-adult transition, and most mutants die during eclosion. Those that do eclose successfully (<10% of pupal) die soon after. In $loqs^{KO}$ homozygous mutants, Loqs proteins were similar to the wild-type levels in the embryos and larvae, while they were not detectable in young adults, suggesting that Loqs proteins are maternally deposited and are depleted during pupal development. Indeed, genetic combinations that remove both maternal and zygotic *loqs* cause death as early as embryonic stage or at first instar larvae, just like *dicer-1* null mutant flies. Thus, *loqs* is crucial for fly development and adult viability. In $loqs^{KO}$ embryos, mature miRNAs decrease and pre-miRNA accumulate for some miRNAs, including miR-125, miR-277, and miR-305. In contrast, for other miRNAs, including miR-1, miR-14, and miR-184, the corresponding pre-miRNA accumulates while mature miRNA levels are unchanged. Finally, for some miRNAs, including *let*-7 and miR-276a, the level of neither mature nor pre-miRNA changes [108]. Thus, different miRNAs have different degrees of dependency on *loqs* for their production.

7.2. *Pact* and *Trbp* mutant phenotypes

PACT knockout mice have defects in ear development and hearing, craniofacial development, growth, and fertility [110]. In humans, PACT mutations (a P222L missense mutation and frame shift mutations causing premature termination) are associated with Dyt16, an autosomal recessive, young-onset dystonia-parkinsonism disorder [111,112]. Dyt16 patients exhibit retarded speech learning in infancy and involuntary muscle contraction beginning as teenagers.

TRBP knockout mice are male sterile and have defects in spermatogenesis [113]. In humans, TRBP mutations (frame shift mutations causing premature termination) are found in sporadic and hereditary carcinomas with microsatellite instability [114]. TRBP, Dicer, and miRNA levels are reduced in the mutant cells. Thus, both PACT and TRBP are important for normal cellular functions in mice, just as *loqs* is in flies.

8. LOQS-PA AND LOQS-PB FUNCTION IN THE miRNA PATHWAY

8.1. Cell culture studies

Multiple lines of evidence show that Loqs-PA and Loqs-PB function directly in the miRNA biogenesis pathway. In cultured *Drosophila* S2 cells in which all isoforms of Loqs proteins were depleted by RNAi, some pre-miRNAs, including those from *bantam*, miR-8, and miR-277, accumulate, just as they do in Dicer-1-depleted cells [22–24]. Moreover, S2 cell lysates depleted of Loqs or Dicer-1 have a reduced ability to convert pre-*let*-7 and pre-*bantam* into mature miRNAs. Dicer-1 and Loqs-PA and Loqs-PB coelute with the peak of pre-miRNA processing activity [22,24], and epitope-tagged Loqs-PA and Loqs-PB coimmunoprecipitate with Dicer-1 and pre-miRNA processing activity [22,24,27]. When FLAG-tagged

Dicer-1 is immunoprecipitated under conditions that dissociate Loqs-PB, pre-miRNA processing activity is reduced; adding purified, recombinant Loqs-PB protein increased pre-miRNA processing [23]. Finally, copurified recombinant Dicer-1/Loqs-PB complex is more active for pre-miRNA processing than Dicer-1 alone [24].

8.2. Kinetic studies using recombinant proteins

Michaelis–Menten analysis of purified, recombinant Dicer-1 and Loqs protein isoforms shows that Loqs-PA increases the affinity of Dicer-1 for pre-miR-305, but not for pre-*let*-7 [28]. In contrast, Loqs-PB increases substrate affinity of Dicer-1 for both pre-miR-305 and pre-*let*-7. In fact, earlier experiments showed that recombinant Dicer-1 bound to pre-*let*-7 more tightly when supplemented with recombinant Loqs-PB [24]. Together, the data establish that Loqs-PB directly enhances pre-miRNA processing by Dicer-1, and that increasing RNA substrate affinity is one of the mechanisms of enhancement. For pre-miR-305, but not for pre-*let*-7, Loqs-PB also increases enzyme turnover, perhaps by facilitating product release.

8.3. *In vivo* fly studies

Studies of *loqs* mutant flies establish that Loqs proteins act in miRNA biogenesis *in vivo*. In $loqs^{f00791}$ hypomorphic mutant flies, mature miRNAs decrease and pre-miRNAs accumulate for *bantam*, miR-7, and miR-277 in male and female whole flies, ovaries, and female carcasses [22,24]. The activity of endogenous miRNA in live flies can be monitored by reporter transgenes bearing miRNA-binding sites in their 3' UTR sequence. The $loqs^{f00791}$ mutant flies show defects in miRNA-mediated repression of green fluorescent protein (GFP) reporters expressed in the eye, while *r2d2* mutants showed no defect in GFP silencing, showing that GFP was suppressed by the miRNA pathway, rather than siRNA pathway [22]. Loqs is required for the production of many, but not all, miRNAs in flies. In $loqs^{KO}$ fly embryos, some mature miRNAs decrease with a corresponding increase in their pre-miRNAs (e.g., miR-125, miR-277, and miR-305), while for other miRNAs, the pre-miRNA accumulates but the level of mature miRNA is unaltered (e.g., miR-1, miR-14, and miR-184). A few miRNAs appear to be Loqs-independent, with neither the mature nor pre-miRNA level changing in the absence of Loqs (e.g., *let*-7 and miR-276a) [108,109]. A Loqs-PB but not Loqs-PA transgene rescued global miRNA and pre-miRNA levels [28,109].

A recent detailed analysis used transgenes to express all viable combinations of the three Loqs isoforms in the $loqs^{KO}$ mutant flies [28]. The lethality of the $loqs^{KO}$ mutant is rescued in adult flies by all combinations of transgenes that expressed Loqs-PA or Loqs-PB, but Loqs-PD alone fails to rescue adult viability. Since *dicer-1* and *ago1* null mutants are lethal, whereas *dicer-2, ago2,* and *r2d2* null mutants are all viable, the lethality caused by the loss of Loqs likely reflects the role of the Loqs-PA and Loqs-PB in the miRNA pathway, rather than the role of Loqs-PD in the siRNA pathway.

The functions of Loqs-PA and Loqs-PB are only partially overlapping: Loqs-PB is required for female fertility and the maintenance of ovarian germline stem cells, as we will discuss in Section 11. High-throughput sequencing of small RNAs from the ovaries and heads of the Loqs isoform-specific rescue flies confirms and extends earlier studies: production of *let-7*, miR-1, miR-184, miR-263a, miR-312, and miR-996 requires Loqs-PA or Loqs-PB, while the abundance of miRNAs such as miR-79, miR-283, miR-305, miR-311, and miR-318 specifically requires Loqs-PB [28].

loqs functions in the miRNA biogenesis pathway, but not in the assembly or function of miRNAs in Ago1-RISC: extracts from $loqs^{KO}$ null mutant egg, can readily be programmed with a synthetic *let-7/let-7★* duplex to form functional *let-7*:Ago1 complexes [108]. Moreover, the amount of miR-14, a Loqs-independent miRNA, that is assembled into Ago1-RISC is unperturbed in $loqs^{KO}$ null mutant eggs. These data suggest that unlike the essential role played by R2D2 in assembling Ago2-RISC, Loqs plays no role in the assembly of Ago1-RISC [16,17].

Thus Loqs proteins, and especially the PB isoform, play a direct and important role in miRNA biogenesis in flies.

9. LOQS-PD FUNCTIONS IN THE siRNA PATHWAY

9.1. Before the discovery of Loqs-PD

In 2008, high-throughput sequencing of small RNAs identified the first fly endo-siRNAs [56–59,115]. endo-siRNAs are genome encoded and derive from esiRNAs, convergent mRNAs (*cis*-NAT endo-siRNAs), or dsRNA from mobile elements. In contrast, exo-siRNAs derive from long dsRNA generated during viral replication [51] and from long dsRNA introduced experimentally or by engineering transgenes that produce long, fully complementary hairpin transcripts [52–54]. Biogenesis of endo-siRNAs depends on *dicer-2* and *loqs* [56–59,115]. Endo-siRNAs, particularly esiRNAs, decrease in S2 cells when all of the *loqs* isoforms, *dicer-2* or *ago2*, but not *r2d2*, *dicer-1*, *ago1*, *pasha*, *drosha*, or *exp5*, are depleted [57,59]. Similarly, endo-siRNAs are decreased in ovaries from $loqs^{f00791}$ hypomorphic or *dicer-2* null mutant flies and in pharate adult flies of $loqs^{KO}$, *dicer-2* null, or *ago2* null mutants.

These findings preceded the discovery of the Loqs-PD isoform [25,26], so it was surprising that *loqs* seemed to function in the siRNA pathway together with Dicer-2. Until then, it was generally believed that Loqs (Loqs-PA and Loqs-PB) and its paralog R2D2 function distinctly in the miRNA and siRNA pathways, respectively, via association with Dicer-1 and Dicer-2, respectively. Nevertheless, it is noteworthy that one of the first *loqs* studies [22] had reported that *loqs* was partly required for efficient gene silencing by exo-siRNAs from a transgene producing an inverted repeat hairpin RNA corresponding to *white* exon 3 [52]. Dicer-2 processes the hairpin into *white* exo-siRNAs [116], which silence *white* mRNA expression, causing the eye to be white or orange instead of the wild-type red. Subsequently, $loqs^{f00791}/loqs^{KO}$ mutant flies were found to produce less *white* exo-siRNAs from the hairpin, as well as less siRNA from experimentally injected long dsRNA [117], suggesting a role for the *loqs* gene in the biogenesis of both endo-siRNAs and exo-siRNAs.

9.2. Discovery of Loqs-PD

9.2.1 A Loqs isoform functions in siRNA production

Efforts to understand the role of *loqs* in the siRNA pathway led to the discovery of Loqs-PD [25,26]. An analysis of *loqs* isoforms identified a novel transcript, *loqs-RD* [26]. When Loqs-PD alone or together with other Loqs isoforms are depleted in S2 cells, esiRNA levels decrease; introducing an RNAi-resistant version of Loqs-PD, but not the other Loqs isoforms, rescues esiRNA production [25,26]. In contrast, depletion of all Loqs isoforms other than Loqs-PD does not affect esiRNA biogenesis. Depletion of Loqs-PD does not affect miRNA levels, while overexpression of Loqs-PD increases only esiRNA accumulation [29].

Dicer-2 immunoprecipitates washed with 1 M NaCl do not contain detectable amounts of Loqs-PD or R2D2; supplementing the immuno-precipitated tagged Dicer-2 with purified recombinant Loqs-PD, but not R2D2, enhances processing of both long dsRNA and pre-esiRNA [27]. *In vivo*, exo-siRNAs, esiRNAs, *cis*-NAT endo-siRNAs, and transposon-derived endo-siRNAs are all reduced in $loqs^{KO}$ null mutant flies lacking Loqs-PD, a defect rescued to normal levels by a Loqs-PD transgene [28]. Clearly, Loqs-PD functions in both the exo- and endo-siRNA pathways.

9.2.2 Kinetic studies using recombinant proteins

Detailed kinetic studies show that both R2D2 and Loqs–PD enhance the binding of Dicer-2 to its substrates. Purified, recombinant Loqs–PD increases the affinity of Dicer-2 for long dsRNA by 14-fold [21] and for pre-esiRNA processing by Dicer-2 by fourfold higher [28]. Similarly, coexpressed and copurified recombinant Dicer-2/R2D2 heterodimer shows a threefold higher affinity for long dsRNA compared with Dicer-2 alone. While both Loqs–PD and R2D2 directly enhance siRNA production by increasing the affinity of Dicer-2 for its substrate, only R2D2 suppresses the inherent ability of Dicer-2 to process pre-miRNA [21].

10. LOQS-PB IS REQUIRED TO MAINTAIN OVARIAN GERMLINE STEM CELLS

10.1. A role for *loqs* in stem cell maintenance

It has been demonstrated that the *loqs*^{f00791} hypomorphic mutant female flies are sterile, lay no eggs, and have abnormally small ovaries, while mutant males show reduced fertility and have testes of normal size [22,24]. *Drosophila* ovaries are composed of 16 ovarioles per ovary, each consisting of a developmentally ordered series of egg chambers and having a tubular shape. The egg chambers are produced continuously by division of germline stem cells within germarium, which is located in the most anterior part of ovarioles. The progression in oocyte development occurs anterior to posterior along the ovarioles. *loqs*^{f00791} mutants have a reduced number of egg chambers, suggesting that the mutants fail to maintain germline stem cells or that the germline stem cells fail to divide [22,24].

   Drosophila oogenesis begins at the anterior tip of the germarium, which houses two or three germline stem cells and its differentiating daughters and granddaughters, the cystoblasts and cystocytes. These go on to form the egg chambers. Germaria from 3- to 4-day-old *loqs* mutant ovaries contain no germline stem cells [22,24,109]. The mutant ovaries contain some late-stage oocytes, suggesting that germline stem cells are initially present, but are lost in the first 2 days of adult life.

10.2. Loqs-PB acts within germline stem cells to ensure their self-renewal

Genetic studies show that Loqs-PB is required to maintain germline stem cells [28,109]. Female fertility is fully rescued by Loqs-PB, but not by Loqs-PA or Loqs-PD. Impaired maintenance of germline stem cells in the *loqs*^{KO}

homozygous or $loqs^{f00791}/loqs^{KO}$ trans-heterozygous mutant ovaries is fully rescued by Loqs-PB, but not by Loqs-PA or Loqs-PD. Thus, Loqs-PB is necessary and sufficient for germline stem cell maintenance in ovaries.

In the germaria of $loqs^{f00791}/loqs^{KO}$ trans-heterozygous mutants, differentiated cystoblasts are often observed at the location of germline stem cells, suggesting that the stem cells are lost by precocious differentiation rather than cell death [109]. Differentiation of germline stem cells into cystoblasts depends on the *bag of marbles* gene [118]. $loqs^{f00791}/loqs^{KO}$; *bag of marbles* double null mutant flies accumulate cystoblast-like germ cell tumors, the *bag of marbles* phenotype, suggesting that loss of *loqs* causes germline stem cells to inappropriately differentiate to cystoblasts without self-renewing [109].

Is Loqs-PB function required within germline stem cells or in the adjacent somatic cells that form the stem cell niche? That is, does Loqs-PB enable germline stem cells to self-renew, allow niche cells to send a self-renewal signal, or both? Liu, McKearin, and coworkers generated flies containing *loqs*-deficient germline stem cell clones surrounded by normal somatic niche cells [109]. Despite the presence of functional Loqs-PB in the somatic cells, the germline stem cells failed to self-renew. Conversely, expressing Loqs-PB in the germline stem cells in a germarium whose somatic terminal filament and cap cells lacked the protein showed that germline stem cell self-renewal does not require Loqs-PB in the somatic niche cells. Thus, Loqs-PB acts intrinsically within the stem cells, but not extrinsically in the surrounding somatic cells, to promote germline stem cell self-renewal.

Maintenance of germline stem cells likely requires Loqs-PB to facilitate the efficient production of miRNAs, since other genes involved in the miRNA pathway such as *dicer-1* and *ago1*, are also required for ovarian germline stem cell maintenance [119,120]. Whether this reflects the general loss of Loqs-PB-dependent miRNAs or just a subset of such miRNAs remains unknown. Loqs-PB also influences the balance of miRNA isoforms (isomirs) for at least three fly miRNAs (discussed below), so germline stem cell maintenance might even require a specific isomir of one or a few miRNAs [28].

11. FOR SOME miRNAS, LOQS-PB TUNES MATURE miRNA LENGTH

11.1. Loqs-PB alters the length of miR-307a, miR-87, and miR-316

Dicer determines the length of a miRNA, but either Dicer or Drosha can determine its 5' end, depending on whether the miRNA resides in the 5' or 3' arm of its pre-miRNA. For a miRNA residing in the 5' arm of pre-miRNA, Drosha cleavage defines the 5′ end and Dicer-1 cleavage defines the 3′ end. Conversely, for miRNA residing in the pre-miRNA 3′ arm, Drosha cleavage defines the 3′ end and Dicer defines the 5′ end. One of the remarkable properties of miRNAs is that only a small patch of the sequence—the "seed" (nucleotides 2–8)—determines which mRNAs the miRNA regulates. Thus, a change in the 5′ end of a miRNA will redefine the seed sequence, changing the repertoire of mRNA targets repressed by the miRNA.

Loqs-PB can alter the site at which Dicer-1 cleaves a pre-miRNA, producing a miRNA isoform of different length and seed sequence compared with those made by Dicer-1 alone or by the Dicer-1/Loqs-PA complex (Fig. 2.4) [28]. High-throughput sequencing of small RNAs from ovaries and heads from $loqs^{KO}$ mutants rescued with Loqs-PA, Loqs-PB, Loqs-PD, or combination of Loqs isoforms identified several miRNAs whose length differed between the rescue flies expressing and the flies missing Loqs-PB.

A 23-nt isomir (miR-307a^{23mer}) is normally the most abundant form of miR-307a in flies with Loqs-PB. In the absence of Loqs-PB, however, the dominant length of miR-307a becomes 21 nt (miR-307a^{21mer}). Since

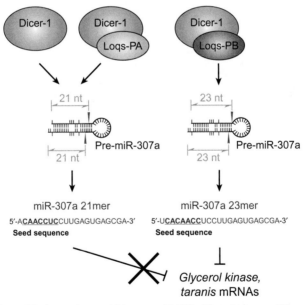

Figure 2.4 Loqs-PB alters where within pre-miR-307a Dicer-1 cleaves. Dicer-1 alone and Dicer-1 associated with Loqs-PA produces 21-nt isoforms of miR-307a and miR-307a*, whereas Dicer-1 bound to Loqs-PB produces 23-nt isoforms. The 21mer and 23mer isomirs of miR-307a have distinct seed sequences (underlined) and thus different mRNA targets. Only the 23mer isomir can repress the *glycerol kinase* and *taranis* mRNAs. (See color plate section in the back of the book.)

miR-307a resides on the 3′ arm of its pre-miRNA, this change in length alters its seed sequence from CACAACC (miR-307a$^{23\text{mer}}$) to CAACCUC (miR-307a$^{21\text{mer}}$). Similarly, miR-307a⋆, which resides on the 5′ arm of its pre-miRNA, becomes shorter in the absence of Loqs-PB.

A similar change in length and seed sequence occurs for miR-87 [28]. A 24-nt isomir of miR-87 (miR-87$^{24\text{mer}}$) is normally the most abundant form of miR-87 in flies with Loqs-PB. In the absence of Loqs-PB, a shorter, 23-nt isomir (miR-87$^{23\text{mer}}$) is more abundant.

In contrast to miR-307a and miR-87 [28], which become shorter in the absence of Loqs-PB, miR-316 becomes longer in the absence of Loqs-PB. A 22-nt isomir of miR-316 is the most dominant isomir in flies with Loqs-PB, while a 23-nt isomir is more abundant in the absence of Loqs-PB. Since miR-316 resides in the 3′ arm of its pre-miRNA, the two miR-316 isomirs share a common seed sequence.

11.2. miR-307a isomirs have distinct target specificities

Both miR-307a$^{21\text{mer}}$ and miR-307a$^{23\text{mer}}$ can silence a partially complementary luciferase reporter in S2 cells, but thus far, target genes have been identified only for miR-307a$^{23\text{mer}}$ [28].

TargetScan [121], a highly validated miRNA target-prediction algorithm, identifies two high confidence targets for miR-307a$^{23\text{mer}}$, *glycerol kinase* (*Gk*) and *taranis* (*tara*). Each of these two genes has single t1A plus 7-nt seed-match target site for 307a$^{23\text{mer}}$ in its 3′ UTR. In cultured S2 cells, transfected miR-307a$^{23\text{mer}}$ but not miR-307a$^{21\text{mer}}$ represses luciferase reporters encompassing the 3′ UTR of each of the genes [28]; when the predicted seed-binding sites are mutated, silencing is lost. Furthermore, endogenous *Gk* and *tara* mRNAs are derepressed in mutant flies lacking Loqs-PB [28]. Therefore, Loqs-PB is required to produce the correct miR-307a isomir to regulate specific mRNAs *in vivo* (Fig. 2.4).

We do not yet know if changes in isomir abundance for the small set of miRNAs like miR-307a contribute to the loss of germline stem cells in ovaries lacking Loqs-PB.

11.3. Loqs-PB acts directly to shift the site at which Dicer-1 cleaves some pre-miRNAs

Biochemical experiments confirm that the altered Dicer-1 cleavage site reflects Loqs-PB binding to Dicer-1. Purified, recombinant Dicer-1 and Dicer-1 supplemented with purified, recombinant Loqs-PA convert pre-miR-307a mainly to miR-307a$^{21\text{mer}}$ (Fig. 2.4). In contrast, Dicer-1 supplemented with Loqs-PB produces significantly more miR-307a^{23mer}. Similar results were obtained for miR-87^{23mer}, while Dicer-1 supplemented with Loqs-PB generates more miR-87^{24mer}. In contrast, Loqs-PB favors production from pre-miR-316 of a shorter, 22-nt isomir of miR-316. Neither Loqs-PA nor Loqs-PB affects where purified Dicer-1 cleaves pre-*let-7*; the length of *let-7* produced *in vivo* is the same in flies expressing only Loqs-PA or Loqs-PB. Mix-and-match experiments in which chimeric pre-miRNAs were created using the stems or loops of pre-*let-7* and pre-miR-307a suggest that the stem of pre-miR-307a is necessary and sufficient to enable Loqs-PB to change where Dicer-1 cleaves [28].

How does Loqs-PB change the nucleotide at which Dicer-1 cleaves pre-miRNA? We propose that the miRNA length produced by Dicer-1 is determined by the distance between the PAZ domain, which recognizes the end of pre-miRNA, and the RNase III domains, which cleave the pre-miRNA (Fig. 2.5). For most pre-miRNAs, Loqs-PB does not change the conformation of their stem, and thus Dicer-1 bound by Loqs-PB produces the same isomir that would be made by Dicer-1 alone. In contrast, Loqs-PB "shrinks" the stems of pre-miR-307a and pre-miR-87, likely by compacting the internal loops. This increases the number of nucleotides that can fit between the Dicer-1 PAZ and RNase III domains, producing a longer isomir. Conversely, Loqs-PB "extends" the stems of a few pre-miRNAs such as pre-miR-316, decreasing the number of nucleotides that fit between the PAZ and RNase III domains, generating a shorter isomir.

11.4. Human and mouse TRBP change the nucleotide at which Dicer cleaves pre-miR-132

Sequencing of small RNAs from immortalized fibroblast cell lines derived from *Trbp* or *Pact* knockout mouse embryos, as well as experiments using purified, recombinant human proteins reveal that human or mouse TRBP, but not PACT, tunes where within pre-miR-132 mammalian Dicer cleaves. Dicer bound to TRBP produces more 22-nt isomir and less 20-nt isomir of miR-132-3p compared with Dicer alone or Dicer bound to PACT [28]. Like the miR-307a isomirs in flies, the miR-132-3p isomirs have distinct seed sequences that would cause them to repress distinct mRNA targets. Similarly, purified human Dicer bound to TRBP generates 1 nt longer isomirs from pre-miR-29, and pre-miR-34c and pre-miR-200a compared with Dicer alone [122].

Thus, Dicer partner proteins in both flies and mammals can change where Dicer cleaves some pre-miRNAs, generating miRNA isoforms with

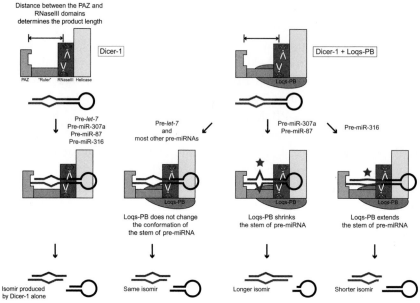

Figure 2.5 A model for how Loqs-PB alters where Dicer-1 cleaves a Pre-miRNA. Dicer-1 product length is determined by the distance between the PAZ and RNase III domains. The number of nucleotides of the pre-miRNA stem that fits that inter-domain distance determines the length of the miRNA and miRNA* products. In the model, Loqs-PB "shrinks" the stems of pre-miR-307a and pre-miR-87, increasing the number of nucleotides that fits and producing a longer isomir. In contrast, Loqs-PB "extends" the stem of pre-miR-316, decreasing the number of nucleotides that fits and generating a shorter isomir. For most pre-miRNAs, Loqs-PB does not alter the conformation of the stem, so it has no effect on product size. (See color plate section in the back of the book.)

distinct lengths, seed sequences, and target specificities compared with those produced by Dicer alone or Dicer bound to alternative dsRBD protein partners.

12. FUTURE DIRECTIONS

Many questions about Dicer partner proteins remain. Why do only some pre-miRNAs require one specific partner rather than another for efficient cleavage by Dicer? Why do some require no partner protein at all? Are there any common structural features among the different classes of pre-miRNAs that reflect which partner protein they require for accurate Dicer processing?

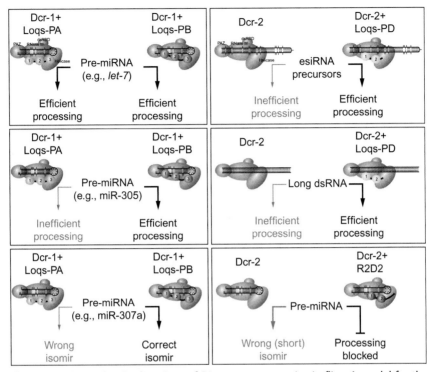

Figure 2.6 The molecular functions of Dicer partner proteins in flies. A model for the functions of R2D2 and the isoforms of Loqs in flies. The numbered spheres denote dsRBDs. (See color plate section in the back of the book.)

Structural studies using crystallography or electron microscopy of Dicer with specific partner proteins and pre-miRNAs may help answer these questions. Loqs-PB differs from Loqs-PA in that it has a longer linker between dsRBD2 and dsRBD3 and its dsRBD3 binds the helicase domain of Dicer-1 more tightly. Since the helicase domain of Dicer-1 binds the terminal loop of pre-miRNA, the first two dsRBDs of Loqs-PB might be less constrained allowing them to reach further along the pre-miRNA stem (Fig. 2.6). For some pre-miRNAs, this could allow Loqs-PB to increase Dicer's substrate affinity and enzyme turnover, and to alter site of cleavage by Dicer-1.

We favor a view that the structural features of the pre-miR307a stem, rather than the miR-307a primary sequence alone, will help explain how Loqs-PB changes where Dicer-1 cuts this pre-miRNA. But it is puzzling that evolution has not selected for "optimized" variants of pre-miR-307a with different miR-307a★ sequences that eliminate the need for Loqs-PB to produce the "right" isomir. We hypothesize that miR-307a^{21mer} plays a distinct function from miR-307a^{23mer} at some stage of development or in a particular cell type. Perhaps the ratio of Loqs-PA versus Loqs-PB is developmentally regulated to ensure the production of the correct relative abundance of isomirs from various pre-miRNAs. In fact, in flies, mice, and humans, the relative abundance of miRNA isomirs generated by distinct Dicer cleavage sites, including 5' isomirs with distinct seed sequences, varies among different tissues and at different developmental stages [123,124].

In this view, the alternative splicing of the *loqs* pre-mRNA must be regulated to achieve the correct isoform ratios in each cell type. How *loqs* alternative splicing is controlled is currently unknown, but is clearly of great interest.

Phosphorylation at four serine residues of TRBP—two in the linker between dsRBD1 and dsRBD2 and two in the linker between dsRBD2 and dsRBD3—stabilizes the Dicer–TRBP complex, altering the global miRNA expression profile in human cells [125]. Serine phosphorylation of two residues in dsRBD3 of PACT increases PACT–PACT homodimerization and PACT binding to protein kinase R (PKR) and decreases binding of PACT to TRBP [126,127]. Similarly, the DGCR8, the human Drosha partner protein, is posttranslationally lysine acetylated, reducing the efficiency with which the Drosha/DGCR8 complex processes pri-miRNA [128]. Histone deacetylase 1 deacetylates the acetylated DGCR8 lysine, restoring the activity of the Drosha/DGCR8 complex. While no posttranslational modification has been reported so far for Loqs, one of the four serine residues phosphorylated in TRBP is conserved in Loqs-PA and Loqs-PB.

Finally, it remains possible that Loqs proteins play additional roles outside the small RNA pathways, since TRBP and PACT both bind dsRNA-dependent PKR inhibiting or enhancing its activity [129,130].

13. CONCLUDING REMARKS

Since the discovery of *loqs* in 2005, our understanding of its molecular functions in flies has increased dramatically, often with important implications for the potential roles of Dicer partner proteins in other animals. Yet, many important and interesting questions remain about Loqs, in particular, and Dicer partner proteins, in general. We look forward to seeing what new and unexpected functions and mechanisms will be discovered in the future.

ACKNOWLEDGMENTS

We thank the members of the Zamore laboratory for help, discussion, advice, and comments on the chapter. We apologize to authors whose work could not be discussed owing to space limitations. This work was supported in part by grants from the National Institutes of Health to P. D. Z. (GM62862 and GM65236), and a JSPS Research Fellowship for Research Abroad and a Charles A. King Trust Postdoctoral Fellowship to R. F.

REFERENCES

[1] Tabara H, Yigit E, Siomi H, Mello CC. The dsRNA binding protein RDE-4 interacts with RDE-1, DCR-1, and a DExH-box helicase to direct RNAi in *C. elegans*. Cell 2002;109:861–71.

[2] Tabara H, et al. The rde-1 gene, RNA interference, and transposon silencing in *C. elegans*. Cell 1999;99:123–32.

[3] Haase AD, et al. TRBP, a regulator of cellular PKR and HIV-1 virus expression, interacts with Dicer and functions in RNA silencing. EMBO Rep 2005;6:961–7.

[4] Chendrimada TP, et al. TRBP recruits the Dicer complex to Ago2 for microRNA processing and gene silencing. Nature 2005;436:740–4.

[5] Chakravarthy S, Sternberg SH, Kellenberger CA, Doudna JA. Substrate-specific kinetics of Dicer-catalyzed RNA processing. J Mol Biol 2010;404:392–402.

[6] MacRae IJ, Ma E, Zhou M, Robinson CV, Doudna JA. In vitro reconstitution of the human RISC-loading complex. Proc Natl Acad Sci USA 2008;105:512–7.

[7] Wang HW, et al. Structural insights into RNA processing by the human RISC-loading complex. Nat Struct Mol Biol 2009;16:1148–53.

[8] Eamens AL, Smith NA, Curtin SJ, Wang MB, Waterhouse PM. The *Arabidopsis thaliana* double-stranded RNA binding protein DRB1 directs guide strand selection from microRNA duplexes. RNA 2009;15:2219–35.

[9] Hiraguri A, et al. Specific interactions between Dicer-like proteins and HYL1/DRB-family dsRNA-binding proteins in *Arabidopsis thaliana*. Plant Mol Biol 2005;57: 173–88.

[10] Vazquez F, Gasciolli V, Crete P, Vaucheret H. The nuclear dsRNA binding protein HYL1 is required for microRNA accumulation and plant development, but not post-transcriptional transgene silencing. Curr Biol 2004;14:346–51.

[11] Han MH, Goud S, Song L, Fedoroff N. The *Arabidopsis* double-stranded RNA-binding protein HYL1 plays a role in microRNA-mediated gene regulation. Proc Natl Acad Sci USA 2004;101:1093–8.

[12] Kurihara Y, Takashi Y, Watanabe Y. The interaction between DCL1 and HYL1 is important for efficient and precise processing of pri-miRNA in plant microRNA biogenesis. RNA 2006;12:206–12.

[13] Dong Z, Han MH, Fedoroff N. The RNA-binding proteins HYL1 and SE promote accurate in vitro processing of pri-miRNA by DCL1. Proc Natl Acad Sci USA 2008;105:9970–5.

[14] Eamens AL, Wook Kim K, Waterhouse PM. DRB2, DRB3 and DRB5 function in a non-canonical microRNA pathway in *Arabidopsis thaliana*. Plant Signal Behav 2012 Oct 1;7(10).

[15] Eamens AL, Kim KW, Curtin SJ, Waterhouse PM. DRB2 is required for microRNA biogenesis in *Arabidopsis thaliana*. PLoS One 2012;7:e35933.

[16] Liu Q, et al. R2D2, a bridge between the initiation and effector steps of the *Drosophila* RNAi pathway. Science 2003;301:1921–5.

[17] Tomari Y, Matranga C, Haley B, Martinez N, Zamore PD. A protein sensor for siRNA asymmetry. Science 2004;306:1377–80.

[18] Tomari Y, Du T, Zamore PD. Sorting of *Drosophila* small silencing RNAs. Cell 2007;130:299–308.

[19] Forstemann K, Horwich MD, Wee L, Tomari Y, Zamore PD. *Drosophila* microRNAs are sorted into functionally distinct Argonaute complexes after production by Dicer-1. Cell 2007;130:287–97.

[20] Okamura K, Robine N, Liu Y, Liu Q, Lai EC. R2D2 organizes small regulatory RNA pathways in *Drosophila*. Mol Cell Biol 2011;31:884–96.

[21] Cenik ES, et al. Phosphate and R2D2 restrict the substrate specificity of Dicer-2, an ATP-driven ribonuclease. Mol Cell 2011;42:172–84.

[22] Forstemann K, et al. Normal microRNA maturation and germ-line stem cell maintenance requires Loquacious, a double-stranded RNA-binding domain protein. PLoS Biol 2005;3:e236.

[23] Saito K, Ishizuka A, Siomi H, Siomi MC. Processing of pre-microRNAs by the Dicer-1-Loquacious complex in *Drosophila* cells. PLoS Biol 2005;3:e235.

[24] Jiang F, et al. Dicer-1 and R3D1-L catalyze microRNA maturation in *Drosophila*. Genes Dev 2005;19:1674–9.

[25] Zhou R, et al. Processing of *Drosophila* endo-siRNAs depends on a specific Loquacious isoform. RNA 2009;15:1886–95.

[26] Hartig JV, Esslinger S, Bottcher R, Saito K, Forstemann K. Endo-siRNAs depend on a new isoform of *loquacious* and target artificially introduced, high-copy sequences. EMBO J 2009;28:2932–44.

[27] Miyoshi K, Miyoshi T, Hartig JV, Siomi H, Siomi MC. Molecular mechanisms that funnel RNA precursors into endogenous small-interfering RNA and microRNA biogenesis pathways in *Drosophila*. RNA 2010;16:506–15.

[28] Fukunaga R, Han BW, Hung JH, Xu J, Weng Z, Zamore PD. Dicer partner proteins tune the length of mature miRNAs in flies and mammals. Cell 2012;151:533–46.

[29] Hartig JV, Forstemann K. Loqs-PD and R2D2 define independent pathways for RISC generation in *Drosophila*. Nucleic Acids Res 2011;39:3836–51.

[30] Lee Y, et al. The nuclear RNase III Drosha initiates microRNA processing. Nature 2003;425:415–9.

[31] Denli AM, Tops BB, Plasterk RH, Ketting RF, Hannon GJ. Processing of primary microRNAs by the microprocessor complex. Nature 2004;432:231–5.

[32] Gregory RI, et al. The microprocessor complex mediates the genesis of microRNAs. Nature 2004;432:235–40.

[33] Han J, et al. The Drosha-DGCR8 complex in primary microRNA processing. Genes Dev 2004;18:3016–27.

[34] Han J, et al. Molecular basis for the recognition of primary microRNAs by the Drosha-DGCR8 complex. Cell 2006;125:887–901.

[35] Okamura K, Hagen JW, Duan H, Tyler DM, Lai EC. The mirtron pathway generates microRNA-class regulatory RNAs in *Drosophila*. Cell 2007;130:89–100.

[36] Ruby JG, Jan CH, Bartel DP. Intronic microRNA precursors that bypass Drosha processing. Nature 2007;448:83–6.

[37] Yi R, Qin Y, Macara IG, Cullen BR. Exportin-5 mediates the nuclear export of pre-microRNAs and short hairpin RNAs. Genes Dev 2003;17:3011–6.

[38] Bohnsack MT, Czaplinski K, Gorlich D. Exportin 5 is a Ran GTP-dependent dsRNA-binding protein that mediates nuclear export of pre-miRNAs. RNA 2004;10:185–91.

[39] Lund E, Guttinger S, Calado A, Dahlberg JE, Kutay U. Nuclear export of microRNA precursors. Science 2004;303:95–8.

[40] Lee YS, et al. Distinct roles for *Drosophila* Dicer-1 and Dicer-2 in the siRNA/miRNA silencing pathways. Cell 2004;117:69–81.

[41] Ghildiyal M, Xu J, Seitz H, Weng Z, Zamore PD. Sorting of *Drosophila* small silencing RNAs partitions microRNA* strands into the RNA interference pathway. RNA 2010;16:43–56.

[42] Czech B, et al. Hierarchical rules for Argonaute loading in *Drosophila*. Mol Cell 2009;36:445–56.

[43] Okamura K, Liu N, Lai EC. Distinct mechanisms for microRNA strand selection by *Drosophila* Argonautes. Mol Cell 2009;36:431–44.

[44] Iki T, et al. In vitro assembly of plant RNA-induced silencing complexes facilitated by molecular chaperone HSP90. Mol Cell 2010;39:282–91.

[45] Iwasaki S, et al. Hsc70/Hsp90 chaperone machinery mediates ATP-dependent RISC loading of small RNA duplexes. Mol Cell 2010;39:292–9.

[46] Miyoshi T, Takeuchi A, Siomi H, Siomi MC. A direct role for Hsp90 in pre-RISC formation in *Drosophila*. Nat Struct Mol Biol 2010;17:1024–6.

[47] Kawamata T, Seitz H, Tomari Y. Structural determinants of miRNAs for RISC loading and slicer-independent unwinding. Nat Struct Mol Biol 2009;16:953–60.

[48] Iwasaki S, Kawamata T, Tomari Y. *Drosophila* Argonaute1 and Argonaute2 employ distinct mechanisms for translational repression. Mol Cell 2009;34:58–67.

[49] Iwasaki S, Tomari Y. Argonaute-mediated translational repression (and activation). Fly (Austin) 2009;3:204–6.

[50] Ghildiyal M, Zamore PD. Small silencing RNAs: an expanding universe. Nat Rev Genet 2009;10:94–108.

[51] Ding S-W. RNA-based antiviral immunity. Nat Rev Immunol 2010;10:632–44.

[52] Lee YS, Carthew RW. Making a better RNAi vector for *Drosophila*: use of intron spacers. Methods 2003;30:322–9.

[53] Kennerdell JR, Carthew RW. Heritable gene silencing in *Drosophila* using double-stranded RNA. Nat Biotechnol 2000;18:896–8.

[54] Kennerdell JR, Carthew RW. Use of dsRNA-mediated genetic interference to demonstrate that *frizzled* and *frizzled 2* act in the Wingless pathway. Cell 1998;95:1017–26.

[55] Yang N, Kazazian HHJ. L1 retrotransposition is suppressed by endogenously encoded small interfering RNAs in human cultured cells. Nat Struct Mol Biol 2006;13:763–71.

[56] Ghildiyal M, et al. Endogenous siRNAs derived from transposons and mRNAs in *Drosophila* somatic cells. Science 2008;320:1077–81.

[57] Czech B, et al. An endogenous small interfering RNA pathway in *Drosophila*. Nature 2008;453:798–802.

[58] Okamura K, Balla S, Martin R, Liu N, Lai EC. Two distinct mechanisms generate endogenous siRNAs from bidirectional transcription in *Drosophila melanogaster*. Nat Struct Mol Biol 2008;15:581–90.

[59] Okamura K, et al. The *Drosophila* hairpin RNA pathway generates endogenous short interfering RNAs. Nature 2008;453:803–6.

[60] Watanabe T, et al. Endogenous siRNAs from naturally formed dsRNAs regulate transcripts in mouse oocytes. Nature 2008;453:539–43.

[61] Tam OH, et al. Pseudogene-derived small interfering RNAs regulate gene expression in mouse oocytes. Nature 2008;453:534–8.

[62] Khvorova A, Reynolds A, Jayasena SD. Functional siRNAs and miRNAs exhibit strand bias. Cell 2003;115:209–16.

[63] Schwarz DS, et al. Asymmetry in the assembly of the RNAi enzyme complex. Cell 2003;115:199–208.

[64] Aza-Blanc P, et al. Identification of modulators of TRAIL-induced apoptosis via RNAi-based phenotypic screening. Mol Cell 2003;12:627–37.

[65] Liu X, Jiang F, Kalidas S, Smith D, Liu Q. Dicer-2 and R2D2 coordinately bind siRNA to promote assembly of the siRISC complexes. RNA 2006;12:1514–20.

[66] Pham JW, Sontheimer EJ. Molecular requirements for RNA-induced silencing com-
     plex assembly in the *Drosophila* RNA interference pathway. J Biol Chem
     2005;280:39278–83.
[67] Liu Y, et al. C3PO, an endoribonuclease that promotes RNAi by facilitating RISC
     activation. Science 2009;325:750–3.
[68] Matranga C, Tomari Y, Shin C, Bartel DP, Zamore PD. Passenger-strand cleavage
     facilitates assembly of siRNA into Ago2-containing RNAi enzyme complexes. Cell
     2005;123:607–20.
[69] Miyoshi K, Tsukumo H, Nagami T, Siomi H, Siomi MC. Slicer function of
     Drosophila Argonautes and its involvement in RISC formation. Genes Dev 2005;
     19:2837–48.
[70] Horwich MD, et al. The *Drosophila* RNA methyltransferase, DmHen1,
     modifies germline piRNAs and single-stranded siRNAs in RISC. Curr Biol
     2007;17:1265–72.
[71] Elbashir SM, Martinez J, Patkaniowska A, Lendeckel W, Tuschl T. Functional anat-
     omy of siRNAs for mediating efficient RNAi in *Drosophila melanogaster* embryo lysate.
     EMBO J 2001;20:6877–88.
[72] Elbashir SM, Lendeckel W, Tuschl T. RNA interference is mediated by 21- and
     22-nucleotide RNAs. Genes Dev 2001;15:188–200.
[73] Zhang H, Kolb FA, Jaskiewicz L, Westhof E, Filipowicz W. Single processing center
     models for human Dicer and bacterial RNase III. Cell 2004;118:57–68.
[74] MacRae IJ, et al. Structural basis for double-stranded RNA processing by Dicer.
     Science 2006;311:195–8.
[75] Basyuk E, Suavet F, Doglio A, Bordonne R, Bertrand E. Human let-7 stem-loop
     precursors harbor features of RNase III cleavage products. Nucleic Acids Res
     2003;31:6593–7.
[76] Cerutti L, Mian N, Bateman A. Domains in gene silencing and cell differentiation pro-
     teins: the novel PAZ domain and redefinition of the Piwi domain. Trends Biochem
     Sci 2000;25:481–2.
[77] Yan KS, et al. Structure and conserved RNA binding of the PAZ domain. Nature
     2003;426:468–74.
[78] Ma JB, Ye K, Patel DJ. Structural basis for overhang-specific small interfering RNA
     recognition by the PAZ domain. Nature 2004;429:318–22.
[79] MacRae IJ, Zhou K, Doudna JA. Structural determinants of RNA recognition and
     cleavage by Dicer. Nat Struct Mol Biol 2007;14:934–40.
[80] Lingel A, Simon B, Izaurralde E, Sattler M. Nucleic acid 3′-end recognition by the
     Argonaute2 PAZ domain. Nat Struct Mol Biol 2004;11:576–7.
[81] Song JJ, et al. The crystal structure of the Argonaute2 PAZ domain reveals an RNA
     binding motif in RNAi effector complexes. Nat Struct Biol 2003;10:1026–32.
[82] Lingel A, Simon B, Izaurralde E, Sattler M. Structure and nucleic-acid binding of the
     Drosophila Argonaute 2 PAZ domain. Nature 2003;426:465–9.
[83] Park JE, et al. Dicer recognizes the 5' end of RNA for efficient and accurate
     processing. Nature 2011;475:201–5.
[84] Zamore PD. Thirty-three years later, a glimpse at the Ribonuclease III active site. Mol
     Cell 2001;8:1158–60.
[85] Takeshita D, et al. Homodimeric structure and double-stranded RNA cleavage
     activity of the C-terminal RNase III domain of human dicer. J Mol Biol
     2007;374:106–20.
[86] Gan J, et al. Structural insight into the mechanism of double-stranded RNA processing
     by ribonuclease III. Cell 2006;124:355–66.
[87] Yamashita S, et al. Structures of the first and second double-stranded RNA-binding
     domains of human TAR RNA-binding protein. Protein Sci 2011;20:118–30.

[88] Yang JS, et al. Conserved vertebrate mir-451 provides a platform for Dicer-independent, Ago2-mediated microRNA biogenesis. Proc Natl Acad Sci USA 2010;107:15163–8.

[89] Sohn SY, et al. Crystal structure of human DGCR8 core. Nat Struct Mol Biol 2007;14:847–53.

[90] Qin H, et al. Structure of the *Arabidopsis thaliana* DCL4 DUF283 domain reveals a noncanonical double-stranded RNA-binding fold for protein-protein interaction. RNA 2010;16:474–81.

[91] Du Z, Lee JK, Tjhen R, Stroud RM, James TL. Structural and biochemical insights into the dicing mechanism of mouse Dicer: a conserved lysine is critical for dsRNA cleavage. Proc Natl Acad Sci USA 2008;105:2391–6.

[92] Mueller GA, et al. Solution structure of the Drosha double-stranded RNA-binding domain. Silence 2010;1:2.

[93] Lau PW, Potter CS, Carragher B, MacRae IJ. Structure of the human Dicer-TRBP complex by electron microscopy. Structure 2009;17:1326–32.

[94] Lau PW, et al. The molecular architecture of human Dicer. Nat Struct Mol Biol 2012;19:436–40.

[95] Ye X, Paroo Z, Liu Q. Functional anatomy of the *Drosophila* microRNA-generating enzyme. J Biol Chem 2007;282:28373–8.

[96] Bianco PR, Kowalczykowski SC. Translocation step size and mechanism of the RecBC DNA helicase. Nature 2000;405:368–72.

[97] Beran RK, Bruno MM, Bowers HA, Jankowsky E, Pyle AM. Robust translocation along a molecular monorail: the NS3 helicase from hepatitis C virus traverses unusually large disruptions in its track. J Mol Biol 2006;358:974–82.

[98] Dumont S, et al. RNA translocation and unwinding mechanism of HCV NS3 helicase and its coordination by ATP. Nature 2006;439:105–8.

[99] Bowers HA, et al. Discriminatory RNP remodeling by the DEAD-box protein DED1. RNA 2006;12:903–12.

[100] Halls C, et al. Involvement of DEAD-box proteins in group I and group II intron splicing. Biochemical characterization of Mss116p, ATP hydrolysis-dependent and -independent mechanisms, and general RNA chaperone activity. J Mol Biol 2007;365:835–55.

[101] Pyle AM. Translocation and unwinding mechanisms of RNA and DNA helicases. Annu Rev Biophys 2008;37:317–36.

[102] Franks TM, Singh G, Lykke-Andersen J. Upf1 ATPase-dependent mRNP disassembly is required for completion of nonsense-mediated mRNA decay. Cell 2010;143:938–50.

[103] Tsutsumi A, Kawamata T, Izumi N, Seitz H, Tomari Y. Recognition of the pre-miRNA structure by *Drosophila* Dicer-1. Nat Struct Mol Biol 2011;18:1153–8.

[104] Welker NC, et al. Dicer's helicase domain discriminates dsRNA termini to promote an altered reaction mode. Mol Cell 2011;41:589–99.

[105] Lee Y, et al. The role of PACT in the RNA silencing pathway. EMBO J 2006;25:522–32.

[106] Moore MJ, Proudfoot NJ. Pre-mRNA processing reaches back to transcription and ahead to translation. Cell 2009;136:688–700.

[107] Lim do H, Kim J, Kim S, Carthew RW, Lee YS. Functional analysis of dicer-2 missense mutations in the siRNA pathway of *Drosophila*. Biochem Biophys Res Commun 2008;371:525–30.

[108] Liu X, et al. Dicer-1, but not Loquacious, is critical for assembly of miRNA-induced silencing complexes. RNA 2007;13:2324–9.

[109] Park JK, Liu X, Strauss TJ, McKearin DM, Liu Q. The miRNA pathway intrinsically controls self-renewal of *Drosophila* germline stem cells. Curr Biol 2007;17:533–8.

[110] Rowe TM, Rizzi M, Hirose K, Peters GA, Sen GC. A role of the double-stranded RNA-binding protein PACT in mouse ear development and hearing. Proc Natl Acad Sci USA 2006;103:5823–8.

[111] Camargos S, et al. DYT16, a novel young-onset dystonia-parkinsonism disorder: identification of a segregating mutation in the stress-response protein PRKRA. Lancet Neurol 2008;7:207–15.

[112] Seibler P, et al. A heterozygous frameshift mutation in PRKRA (DYT16) associated with generalised dystonia in a German patient. Lancet Neurol 2008;7:380–1.

[113] Zhong J, Peters AH, Lee K, Braun RE. A double-stranded RNA binding protein required for activation of repressed messages in mammalian germ cells. Nat Genet 1999;22:171–4.

[114] Melo SA, et al. A TARBP2 mutation in human cancer impairs microRNA processing and DICER1 function. Nat Genet 2009;41:365–70.

[115] Kawamura Y, et al. *Drosophila* endogenous small RNAs bind to Argonaute2 in somatic cells. Nature 2008;453:793–7.

[116] Vagin VV, et al. A distinct small RNA pathway silences selfish genetic elements in the germline. Science 2006;313:320–4.

[117] Marques JT, et al. Loqs and R2D2 act sequentially in the siRNA pathway in *Drosophila*. Nat Struct Mol Biol 2010;17:24–30.

[118] Ohlstein B, McKearin D. Ectopic expression of the *Drosophila* Bam protein eliminates oogenic germline stem cells. Development 1997;124:3651–62.

[119] Yang B, et al. The muscle-specific microRNA miR-1 regulates cardiac arrhythmogenic potential by targeting GJA1 and KCNJ2. Nat Med 2007;13:486–91.

[120] Jin Z, Xie T. Dcr-1 maintains *Drosophila* ovarian stem cells. Curr Biol 2007;17:539–44.

[121] Lewis BP, Shih IH, Jones-Rhoades MW, Bartel DP, Burge CB. Prediction of mammalian microRNA targets. Cell 2003;115:787–98.

[122] Lee HY, Doudna JA. TRBP alters human precursor microRNA processing in vitro. RNA. 2012 Nov;18(11):2012-9.

[123] Fernandez-Valverde SL, Taft RJ, Mattick JS. Dynamic isomiR regulation in *Drosophila* development. RNA 2010;16:1881–8.

[124] Lee LW, et al. Complexity of the microRNA repertoire revealed by next-generation sequencing. RNA 2010;16:2170–80.

[125] Paroo Z, Ye X, Chen S, Liu Q. Phosphorylation of the human microRNA-generating complex mediates MAPK/Erk signaling. Cell 2009;139:112–22.

[126] Peters GA, Li S, Sen GC. Phosphorylation of specific serine residues in the PKR activation domain of PACT is essential for its ability to mediate apoptosis. J Biol Chem 2006;281:35129–36.

[127] Singh M, Patel RC. Increased interaction between PACT molecules in response to stress signals is required for PKR activation. J Cell Biochem 2012;113:2754–64.

[128] Wada T, Kikuchi J, Furukawa Y. Histone deacetylase 1 enhances microRNA processing via deacetylation of DGCR8. EMBO Rep 2012;13:142–9.

[129] Patel CV, Handy I, Goldsmith T, Patel RC. PACT, a stress-modulated cellular activator of interferon-induced double-stranded RNA-activated protein kinase, PKR. J Biol Chem 2000;275:37993–8.

[130] Daher A, et al. Two dimerization domains in the trans-activation response RNA-binding protein (TRBP) individually reverse the protein kinase R inhibition of HIV-1 long terminal repeat expression. J Biol Chem 2001;276:33899–905.

CHAPTER THREE

Translin:TRAX Complex (C3PO), a Novel Ribonuclease for the Degradation of Small RNAs

Hong Zhang[*,†,1], **Qinghua Liu**[†,1]

[*]Department of Biophysics, University of Texas Southwestern Medical Center, Dallas, Texas, USA
[†]Department Biochemistry, University of Texas Southwestern Medical Center, Dallas, Texas, USA
[1]Corresponding authors: e-mail address: zhang@chop.swmed.edu; Qinghua.Liu@UTSouthwestern.edu

Contents

Abstract

C3PO is a heteromeric complex of two evolutionarily related proteins, translin and TRAX. C3PO was recently discovered to possess ribonuclease activity and promote the activation of RNA-induced silencing complex (RISC) in *Drosophila* and human by degrading the passenger strand fragments of duplex siRNA. C3PO is also involved in tRNA processing in the filamentous fungus *Neurospora crassa* and degrades the 5′-leader fragment that is cleaved off the precursor tRNA (pre-tRNA) by RNase P. Structural and biochemical studies revealed that active C3PO adopts a football-shaped hetero-octamer configuration containing six translin and two TRAX subunits. The RNA binding and catalytic residues are located at the interface between TRAX and translin subunits facing the hollow interior of C3PO octamer, indicating that C3PO binds and cleaves RNA inside its largely closed barrel. The ability of translin and TRAX to form complexes of different oligomeric states, such as tetramers, hexamers, and octamers, suggests a novel mechanism for the recruitment of RNA substrate and the assembly of active C3PO. Further mechanistic and functional studies are required to fully understand the role of C3PO in a number of biological processes.

The Enzymes, Volume 32
ISSN 1874-6047
http://dx.doi.org/10.1016/B978-0-12-404741-9.00003-9

1. INTRODUCTION: EARLIER STUDIES OF TRANSLIN AND TRAX

Translin, also known as testis and brain-specific RNA-binding protein (TB-RBP), was independently discovered in several laboratories in the 1990s [1–4]. Kasai *et al.* first identified a 27 kDa protein that avidly bound to the consensus DNA sequences at the breakpoint junctions of chromosomal translocations in human lymphoma, and they name the protein translin to signify its possible involvement in chromosomal translocations [1]. Hecht *et al.* found that translin was part of a protein complex that binds to the conserved elements in the 3′-untranslated regions of specific mRNAs in spermatocytes, mediating the delay in translation [5]. Additionally, two other groups found that translin was part of a single-stranded DNA (ssDNA) binding complex enriched in brain that has high affinity for G-rich sequences [3,4,6]. Subsequently, another 33 kDa translin-like protein was identified as a translin–interacting protein by yeast two-hybrid screen and was named translin-associated factor X (TRAX) [7]. Translin and TRAX are broadly expressed in multiple tissues such as brain, testes, kidney, liver, and hemopoietic cells [2,5,8–10], and they are evolutionarily conserved from fission yeast to human, suggesting that they may play fundamental roles in cellular processes. Although homologs of Translin are also found in some bacteria and archaea, their function in the prokaryotes has not been characterized.

Multiple lines of evidence suggest that translin and TRAX proteins form a tight complex and endogenous translin and TRAX heteromeric complex is likely the predominant functional species of the proteins *in vivo* [11–14]. In particular, TRAX protein is unstable on its own and its stability depends on its association with translin both *in vitro* and *in vivo*. Genetic deletion of *translin* also diminishes the level of TRAX protein in yeast, *Drosophila*, and mice [11,13–15]. Moreover, biochemical fractionation of *Drosophila* and mammalian cell extracts revealed that the translin and TRAX exist as a complex with a specific stoichiometry [10,16,17] (Fig. 3.1), further supporting that translin:TRAX complex (Renamed C3PO. See next section.) is the dominant functional species in cells. Nevertheless, translin by itself forms stable homo-oligomers and may be present in certain tissues or cell types.

Translin and TRAX knockout mice and mutant flies are viable but are growth retarded at birth and display a number of fertility and behavior

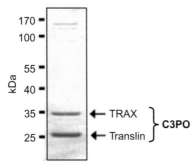

170 -
100 -

55 -
kDa 40 -

35 -   ← TRAX }
                          } C3PO
25 -   ← Translin }

Figure 3.1 Colloidal blue-stained SDS-PAGE of purified endogenous human C3PO from HeLa cells. Note that the intensities of the translin and TRAX bands indicate a 6:2 stoichiometry of the complex.

abnormalities [11,18–20]. Microarray analysis identified altered gene expression patterns in these translin knockout mice [11]. Several studies also suggest that C3PO mediates dendritic trafficking of RNAs, a process believed to play a critical role in synaptic plasticity [19,21–23]. However, the biochemical mechanism underlying the diverse cellular functions of translin and TRAX was poorly understood. Previously, C3PO has been known to possess ssRNA- and ssDNA-binding activity. Translin has a nuclear export signal, whereas TRAX has a nuclear localization signal; C3PO has been shown to shuttle between cytoplasm and nucleus [24]. There are many gaps in our knowledge of the molecular mechanisms underlying C3PO function. The recent discovery that C3PO but not translin itself possesses potent ribonuclease activity sheds important new lights onto these enigmatic proteins [16]. In this chapter, we summarize the results from the recently biochemical and structural studies of C3PO and its roles in RNA interference (RNAi) and tRNA processing [16,17,25,26].

2. DISCOVERY OF C3PO AS A NOVEL ENDORIBONUCLEASE INVOLVED IN THE ACTIVATION OF RNA-INDUCED SILENCING COMPLEX

RNAi is an evolutionarily conserved, double-stranded RNA (dsRNA)-induced posttranscriptional gene-silencing mechanism [27,28]. Typically, long dsRNA molecules are processed by ribonuclease Dicer into ~21-nt small interfering RNA (siRNA), which directs sequence-specific cleavage of complementary mRNA [29–31]. The exogenous siRNAs function as a potent defense mechanism against invading nucleic acids, such as viruses and transgenes [32]. On the other hand, endogenous (endo)-siRNAs, which can be generated from transposable elements, complementary annealed transcripts, and long fold-back transcripts, play important roles in diverse biological processes, including heterochromatin regulation in yeast and transposon silencing in *Drosophila* [33].

The effector complex of RNAi is the RNA-induced silencing complex (RISC), wherein single-stranded siRNA directs Argonaute (e.g., Ago2) endonuclease to catalyze sequence-specific target mRNA cleavage. Assembly of active RISC consists of loading duplex siRNA (passenger strand–guide strand) onto Ago2 and subsequent removal of the passenger strand [34–38] (Fig. 3.2). In *Drosophila*, Dicer-2 (Dcr-2) and its dsRNA-binding partner R2D2 are required to recruit duplex siRNA to Ago2 to promote RISC assembly [30,31,39,40]. The Dcr-2/R2D2 complex is the core component of the RISC-loading complex that differentiates the two strands of siRNA by sensing the thermodynamic asymmetry of siRNA ends [41]. In addition, Ago2-associated heat shock proteins have also been shown to promote siRNA loading onto Ago2 [42–44]. After

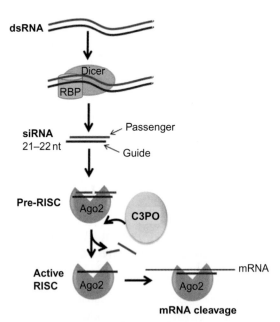

Figure 3.2 RNAi pathway and the role of C3PO in RISC activation. RBP, RNA-binding protein, for example, R2D2. (For color version of this figure, the reader is referred to the online version of this chapter.)

duplex siRNA is recruited to Ago2, the endonuclease activity of Ago2 nicks the passenger strand into a 9-nt and a 12-nt fragment [34–38]. Recombinant *Drosophila* Dcr-2/R2D2 and Ago2 proteins can reconstitute a basal level of duplex siRNA-initiated RISC activity. This core reconstitution system was used to purify a novel RISC-enhancing factor from *Drosophila* S2 cell extract by chromatographic fractionation, which turned out to be a complex of translin and TRAX [16]. This complex was hence named C3PO (*component 3 promoter of* RISC) because, in addition to Dcr-2 and R2D2, it was the third component that promotes the RISC activity.

It has also been demonstrated that RNAi is less efficient in *translin* (*trsn*) mutant *Drosophila* embryos that lack both translin and TRAX proteins [16]. Consistently, *trsn* mutant ovary extract exhibits ∼25% of the duplex siRNA-initiated RISC activity of wild-type extract, and this defect can be rescued by addition of recombinant C3PO. The inefficient RNAi activity was accompanied by increased accumulation of both 9-nt and 12-nt passenger strand fragments in *trsn* mutant cell extract. Addition of C3PO resulted in rapid degradation of the siRNA passenger fragments in the Dcr-2/R2D2/Ago2 reconstitution system, indicating for the first time that C3PO is a ribonuclease [16]. Further biochemical characterization revealed that C3PO has potent Mg^{2+}-dependent nuclease activity toward ssRNA, but not dsRNA. C3PO cleaves both linear and circular ssRNA almost equally efficient, indicating that it is an endoribonuclease [16]. Bioinformatic and mutagenesis studies led to the identification of Glu123, Glu126, and Asp204 on *Drosophila* TRAX to be the critical catalytic residues, demonstrating that TRAX is the catalytic subunit of C3PO. Mutation of each of these residues to alanine abolished the RNase and the RISC-enhancing activities of C3PO, suggesting that the RNase activity of C3PO is required for its ability to activate RISC assembly. Subsequent studies showed that human C3PO plays a similar role in RISC activation as *Drosophila* C3PO, degrading the nicked passenger strand fragments [17]. Moreover, recombinant human Ago2 and C3PO can reconstitute robust duplex siRNA-initiated RISC activity in the absence of Dicer–TRBP complex, suggesting a Dicer-independent mechanism for human Ago2-RISC activation [17].

3. C3PO RIBONUCLEASE IN tRNA PROCESSING

Translin and TRAX are evolutionarily conserved from fission yeast (*Schizosaccharomyces pombe*) to human, though *Saccharomyces cerevisiae* (budding yeast) is a notable exception, lacking both translin and TRAX-encoding genes as well as other components of the RNAi machinery. In the filamentous fungus *Neurospora crassa*, a model eukaryotic system for RNAi studies, the exonuclease QIP interacts with the Argonaute protein QDE-2 and plays a similar role as C3PO in degrading the nicked passenger strand of duplex siRNA to activate RISC [45]. A recent study reported that *Neurospora* C3PO (nC3PO) does not play a significant role in RNAi in this fungus [26]. It was observed that a ladder of ∼18–50 nt small RNA (sRNA) species accumulated to high levels in both *trsn* and *trx* knockout strains and the most abundant species are the ∼20-nt 5′-leader fragments of precursor tRNA (pre-tRNA) [26]. This study established that nC3PO is involved in the maturation of tRNA and functions as an RNase degrading the 5′-leader RNA after it is cleaved by RNase P from pre-tRNA. In the mammalian system, levels of several tRNAs are significantly increased in *translin* knockout ($trsn^{-/-}$) mouse embryonic fibroblast (MEF) compared to the wild-type MEFs. Additionally, a ∼25-nt tRNA-derived fragment was missing in the $trsn^{-/-}$ cells, suggesting that mammalian C3PO is also involved in tRNA processing. In addition to pre-tRNA processing, sRNA accumulation in $trsn^{-/-}$ cells was found for 18 other non-tRNA loci, suggesting that nC3PO has other non-tRNA substrates.

4. STRUCTURE OF C3PO: AN UNUSUAL ASYMMETRICAL HETERO-OCTAMER

Following the discovery that C3PO is an RNase involved in RISC activation, the crystal structures of C3POs were determined independently in three different laboratories [17,25,46]. The structure of full-length human C3PO revealed a hetero-octameric assembly shaped like a football with a large central cavity [17] (Fig. 3.3A). The C3PO hetero-octamer has an unusual asymmetric stoichiometry of two TRAX and six translin, similar to that found in the purified endogenous C3PO (Fig. 3.1) [10,16,17]. The TRAX subunit has a similar three-dimensional structure to that of translin [47,48] and contains seven α-helices with six of them forming a two-layered right-handed superhelical bundle (Fig. 3.3B). In C3PO, each TRAX subunit heterodimerizes with a translin subunit (Fig. 3.3C), and the whole C3PO assembly consists of two translin:TRAX heterodimers and two translin:translin homodimers. A close inspection of the unusual asymmetric arrangement of the subunits revealed that the four dimers of the octamer display a superhelical shift along the long axis of the football-shaped barrel, leading to the breakdown of the pseudo four-fold

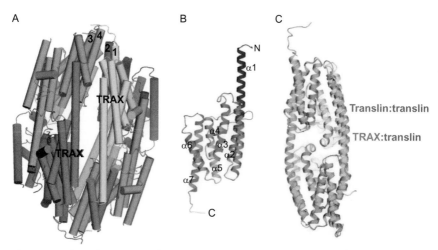

Figure 3.3 Crystal structure of human C3PO. (A) A cylindrical cartoon presentation of the hetero-octamer structure of human C3PO. The two TRAX subunits are colored yellow and orange, whereas the translin subunits are colored cyan, green, and dark green. (B) Ribbon diagram of TRAX monomer showing the right-handed α/α superhelical fold. The figure is color ramped from blue (N-terminus) to red (C-terminus). (C) Superposition of translin homodimer (gray) on translin (cyan):TRAX (yellow) heterodimer. The similar dimer conformation observed in all C3PO and translin structures obtained so far suggests that these dimers are the basic building blocks of C3PO. (See color plate section in the back of the book.)

rotational symmetry of the complex (Fig. 3.3A). A similar superhelical shift also exists among the subunits in the translin homo–octamer structures determined earlier [47,48]. Therefore, it is likely that this asymmetric spiral arrangement of the subunits reflects an intrinsic structural property of translin and TRAX and leads to the asymmetric incorporation of two TRAX in C3PO octamer.

In the crystal structure of C3PO, the conserved catalytic residues on TRAX subunit, including Glu126, Glu129, Asp193 identified previously and Glu197 based on the human C3PO:Mn^{2+} structure [17], are located in close proximity facing the hollow interior of the C3PO barrel. The structure of human C3PO complexed to Mn^{2+} ion showed that residues Glu129 and Glu197 of TRAX directly coordinate the metal ion, while the other two catalytic residues Asp193 and Glu126 could either coordinate metal ion indirectly through a water molecule or act as the catalytic base (Fig. 3.4B). The fortuitous presence of four sulfate or phosphate molecules near the Mn^{2+} ion in the crystal structure allowed modeling of a short stretch A-form ssRNA

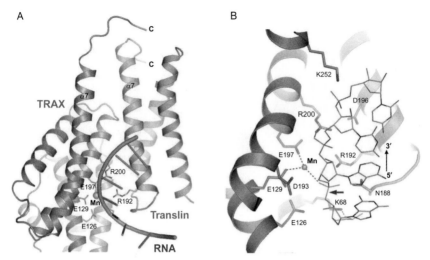

Figure 3.4 RNA binding and catalytic site of C3PO. (A) Cartoon model of an ssRNA (green) in A-form conformation binding at an interface between TRAX (orange) and translin. The Mn^{2+} ion is shown as a magenta sphere. The side chains of key protein residues are shown as sticks. (B) A detailed view of the proposed ssRNA binding and catalytic site. The red arrow points to the phosphoester bond that is cleaved to generate 5′-phosphate and 3′-hydroxyl products. Dotted lines indicate that residues E129 and E197 of TRAX and an ssRNA phosphate are directly coordinated to the catalytic metal ion. (See color plate section in the back of the book.)

by superimposing RNA backbone phosphates onto the SO_4/PO_4 molecules. This model revealed that several positively charged residues nearby could potentially interact with RNA backbone. Mutagenesis results confirmed these predictions and showed that, in addition to the four acidic residues (E126, E129, D193, and E197) that are critical for catalysis, three positively charged residues, Lys68 and Arg200 of TRAX and Arg192 of translin are important for ssRNA substrate binding (Fig. 3.4B). Mutation of each of these residues to alanine diminished ssRNA binding and abolished RNase activity of C3PO [17]. These results validated the predicted nucleotide-binding site, which is located at an interface between the TRAX and the adjacent translin subunit with the largest vertical shift. Therefore, it appears that the specific asymmetric arrangement of the subunits in C3PO hetero-octamer is important for creating a suitable nucleotide substrate-binding site for the catalysis. It is conceivable that a slightly different positioning of these TRAX and translin subunits would distort the substrate-binding groove and compromise the substrate binding and cata-lytic activity of C3PO. These structural and mutagenesis results revealed that C3PO must sequester the ssRNA substrate inside its largely secluded central cavity for cleavage, a process that can be regulated to prevent non-specific RNA degradation. These unique features of C3PO should dictate its biological functionality. To our knowledge, no other nucleases use a similar mechanism for interacting with its substrates.

5. OLIGOMERIC STATES OF C3PO AND IMPLICATIONS FOR SUBSTRATE BINDING AND ACTIVATION

Consistent with the crystallographic observations, solution studies of the nucleic acid binding and catalytic activity of C3PO by size-exclusion chroma-tography and nondenaturing mass spectral analysis have also established that the active C3PO is octameric [17,25,49]. More specifically, nondenaturing mass spectral analysis of full-length recombinant *Drosophila* C3PO showed that the major species of active C3PO octamer contains six translin and two TRAX, consistent with the crystal structure of full-length human C3PO [25]. The largely closed cage-like structure of C3PO octamer with an interior active site immediately raised the question as to how ssRNA substrate is transported inside the barrel. Modeling studies indicated that the interior of C3PO is large enough (maximum height about 70 Å and diameter 40 Å) to accommodate its ssRNA substrates and even double-stranded nucleic acid of limited length. In theory, there could be two scenarios in which ssRNA can reach the active site of C3PO. First, local conformational flexibility of C3PO subunits may lead to transient opening of C3PO barrel wall which would allow ssRNA substrate to enter the barrel. Second, the oligomerization state of C3PO may be dynamic, where lower order translin and TRAX oligomers may coexist with C3PO octamer in equilibrium. The additional crystal structures of a full-length and a truncated construct of *Drosophila melanogaster* C3PO (dC3PO) reported recently support the second scenario [25,46]. These structures revealed different oligomeric states of the C3PO. In the crystal structure of dC3PO determined in Dinshaw Patel's laboratory, where residues 218–235 of translin and residues 1–29 of TRAX were deleted in the protein expression construct in an attempt to facilitate crystallization, the complex is hexameric consisting of two translin:TRAX heterodimers sandwiching a translin homodimer in a bowl-shaped assembly [25]. On the other hand, the full-length dC3PO structure revealed to be a tetramer containing two translin:TRAX heterodimer [46]. In all these C3PO structures as well as previously determined translin homomultimer structures [47,48], the conformations of translin homodimer and translin:TRAX heterodimer are similar to each other, suggesting that these dimers are very conserved and likely the smallest building blocks of various C3POs. Different combinations of these dimers could exist in solution, and the assembly of the active 6:2 hetero–octamer might be facilitated by factors such as the levels of translin and TRAX subunits, and the presence of ssRNA substrates.

It is worth noting that crystal structures of mouse and human translin have been determined previously, and they adopt a similar octameric architecture as that observed for full-length human C3PO, with an asymmetrical spatial arrangement that breaks down the pseudo four-fold symmetry [47,48]. A mutant *Drosophila* translin structure, on the other hand, adopts an open tetrameric conformation [49]. There seems to be little doubt that translin and TRAX homo- and heterodimers could assemble in a number of different ways into different oligomeric states (various tetramers, hexamers, and octamers of different stoichiometries). It is tempting to speculate that the assembly of the active 6:2 C3PO hetero–octamer may be a dynamic process regulated by interactions with ssRNA substrates or potentially other proteins. The details of this process are currently not known.

6. ENZYMATIC PROPERTIES OF C3PO

It was completely unexpected that C3PO is a ribonuclease as no other known DNA or RNA nucleases adopt the similar α/α superhelical fold as translin and TRAX. Nevertheless, biochemical instinct and the observation of sRNAs accumulating in *trsn* and *trx* knockout cells led to the discovery that C3PO indeed possesses potent ribonuclease activity, and its catalytic activity is necessary for its ability to promote activation of RISC and tRNA processing [16,26]. Mutagenesis analysis so far identified four catalytically important residues on TRAX and established that TRAX is the main catalytic subunit of C3PO. C3PO is a metal-dependent endoribonuclease, able to cleave both linear and circular ssRNA efficiently [17,25]. A panel of divalent metal ions (Mg^{2+}, Mn^{2+}, Ni^{2+}, and Zn^{2+}, but not Ca^{2+}) can support catalysis by C3PO. β-Elimination and alkaline phosphate treatment experiments established that the products of C3PO contain a $5'$-phosphate and $2',3'$-hydroxyl groups. Therefore, the cleavage would involve a water molecule for the hydrolytic reaction.

C3PO has a maximum length limit for ssRNA substrates of <80-nt (Q. Liu, unpublished results). And for RNA shorter than 7 nt, the rate of cleavage drops significantly [25]. Preliminary kinetic measurements revealed that the turn-over rate of *Drosophila* C3PO is 0.74 s^{-1} at a high RNA concentration (1.5 μM) using a 21-nt ssRNA substrate [25]. C3PO appears to cleave ssRNA substrates processively leading to single nucleotide products. Although C3PO seems to cleave ssRNA in a largely sequence-independent manner, it does show a certain preference for guanine nucleotides.

The structure of human C3PO in complex with Mn^{2+} ion and the modeling of ssRNA substrate at the active site provided further insight into the catalytic center of C3PO. Specifically, it established that the substrate binding and catalytic site is located at an interface between TRAX and translin. While the majority of catalytic residues identified so far are located on TRAX, the adjacent translin subunit also contributes to ssRNA substrate binding and is critical for the catalysis. The enzymatic properties of C3PO underlie its biological functions and are very much in need of further investigation and characterizations. A full characterization of the catalytic mechanism of this novel ribonuclease would also require determining the complex structures of C3PO with its ssRNA substrate.

7. CONCLUDING REMARKS

Significant progresses have been made in the structural and functional characterizations of C3PO. The discovery that C3PO is a ribonuclease provided a biochemical basis for its newly characterized roles in RISC activation, tRNA processing, and perhaps other cellular functions. C3PO represents a completely new nuclease fold and has unique properties with regard to its active form assembly, substrate recognition and recruitment, nucleotide preferences, and processivity. All these aspects of C3PO function are still at the early stages of investigation. The active form of C3PO is a largely closed asymmetric hetero-octamer barrel with active site located inside its hollow interior. This configuration dictates the length limits of its RNA substrates. C3PO is involved in RISC activation in animals and also plays a conserved role in tRNA processing in both fungi and animals. It is likely that C3PO is also involved in other RNA metabolic processes by degrading sRNA molecules. C3PO is highly conserved from fission yeast to human except in *S. cerevisiae*, suggesting that presence of C3PO may bestow an evolutionary advantage to the organism under certain conditions.

Future studies are needed to fully understand the biochemical and cellular functions of this unique ribonuclease.

ACKNOWLEDGMENTS

We thank members of our laboratories who made important contributions to the studies of C3PO, in particular, Drs. Ying Liu, Nian Huang, and Xuecheng Ye. We also thank Drs. Yi Liu, Qiu-Xing Jiang, and Nick Grishin for discussions, and Dr. Lisa Kinch for critical reading of the chapter. The work was supported by grants from the Welch foundation (I-1608) and National Institute of Health (GM084010 and GM091286) to Q. L., and an American Heart Association grant (10GRNT4310090) to H. Z.

REFERENCES

[1] Aoki K, Suzuki K, Sugano T, Tasaka T, Nakahara K, Kuge O, et al. A novel gene, Translin, encodes a recombination hotspot binding protein associated with chromosomal translocations. Nat Genet 1995;10:167–74.

[2] Han JR, Gu W, Hecht NB. Testis-brain RNA-binding protein, a testicular translational regulatory RNA-binding protein, is present in the brain and binds to the $3'$ untranslated regions of transported brain mRNAs. Biol Reprod 1995;53:707–17.

[3] Taira E, Finkenstadt PM, Baraban JM. Identification of translin and trax as components of the GS1 strand-specific DNA binding complex enriched in brain. J Neurochem 1998;71:471–7.

[4] Aharoni A, Baran N, Manor H. Characterization of a multisubunit human protein which selectively binds single stranded d(GA)n and d(GT)n sequence repeats in DNA. Nucleic Acids Res 1993;21:5221–8.

[5] Wu XQ, Gu W, Meng X, Hecht NB. The RNA-binding protein, TB-RBP, is the mouse homologue of translin, a recombination protein associated with chromosomal translocations. Proc Natl Acad Sci USA 1997;94:5640–5.

[6] Jacob E, Pucshansky L, Zeruya E, Baran N, Manor H. The human protein translin specifically binds single-stranded microsatellite repeats, d(GT)n, and G-strand telomeric repeats, d(TTAGGG)n: a study of the binding parameters. J Mol Biol 2004;344:939–50.

[7] Aoki K, Ishida R, Kasai M. Isolation and characterization of a cDNA encoding a Translin-like protein, TRAX. FEBS Lett 1997;401:109–12.

[8] Jaendling A, McFarlane RJ. Biological roles of translin and translin-associated factor-X: RNA metabolism comes to the fore. Biochem J 2010;429:225–34.

[9] Kwon YK, Hecht NB. Cytoplasmic protein binding to highly conserved sequences in the 3' untranslated region of mouse protamine 2 mRNA, a translationally regulated transcript of male germ cells. Proc Natl Acad Sci USA 1991;88:3584–8.

[10] Wu RF, Osatomi K, Terada LS, Uyeda K. Identification of Translin/TRAX complex as a glucose response element binding protein in liver. Biochim Biophys Acta 2003;1624:29–35.

[11] Chennathukuzhi V, Stein JM, Abel T, Donlon S, Yang S, Miller JP, et al. Mice deficient for testis-brain RNA-binding protein exhibit a coordinate loss of TRAX, reduced fertility, altered gene expression in the brain, and behavioral changes. Mol Cell Biol 2003;23:6419–34.

[12] Li Z, Baraban JM. High affinity binding of the Translin/TRAX complex to RNA does not require the presence of Y or H elements. Brain Res Mol Brain Res 2004;120:123–9.

[13] Yang S, Cho YS, Chennathukuzhi VM, Underkoffler LA, Loomes K, Hecht NB. Translin-associated factor X is post-transcriptionally regulated by its partner protein

TB-RBP, and both are essential for normal cell proliferation. J Biol Chem 2004; 279:12605–14.

[14] Claussen M, Koch R, Jin ZY, Suter B. Functional characterization of *Drosophila* Translin and TRAX. Genetics 2006;174:1337–47.

[15] Jaendling A, Ramayah S, Pryce DW, McFarlane RJ. Functional characterisation of the *Schizosaccharomyces pombe* homologue of the leukaemia-associated translocation breakpoint binding protein translin and its binding partner, TRAX. Biochim Biophys Acta 2008;1783:203–13.

[16] Liu Y, Ye X, Jiang F, Liang C, Chen D, Peng J, et al. C3PO, an endoribonuclease that promotes RNAi by facilitating RISC activation. Science 2009;325:750–3.

[17] Ye X, Huang N, Liu Y, Paroo Z, Huerta C, Li P, et al. Structure of C3PO and mechanism of human RISC activation. Nat Struct Mol Biol 2011;18:650–7.

[18] Stein JM, Bergman W, Fang Y, Davison L, Brensinger C, Robinson MB, et al. Behavioral and neurochemical alterations in mice lacking the RNA-binding protein translin. J Neurosci 2006;26:2184–96.

[19] Li Z, Wu Y, Baraban JM. The Translin/TRAX RNA binding complex: clues to function in the nervous system. Biochim Biophys Acta 2008;1779:479–85.

[20] Suseendranathan K, Sengupta K, Rikhy R, D'Souza JS, Kokkanti M, Kulkarni MG, et al. Expression pattern of *Drosophila* translin and behavioral analyses of the mutant. Eur J Cell Biol 2007;86:173–86.

[21] Kobayashi S, Takashima A, Anzai K. The dendritic translocation of translin protein in the form of BC1 RNA protein particles in developing rat hippocampal neurons in primary culture. Biochem Biophys Res Commun 1998;253:448–53.

[22] Finkenstadt PM, Kang WS, Jeon M, Taira E, Tang W, Baraban JM. Somatodendritic localization of Translin, a component of the Translin/TRAX RNA binding complex. J Neurochem 2000;75:1754–62.

[23] Chiaruttini C, Vicario A, Li Z, Baj G, Braiuca P, Wu Y, et al. Dendritic trafficking of BDNF mRNA is mediated by translin and blocked by the G196A (Val66Met) mutation. Proc Natl Acad Sci USA 2009;106:16481–6.

[24] Cho YS, Chennathukuzhi VM, Handel MA, Eppig J, Hecht NB. The relative levels of translin-associated factor X (TRAX) and testis brain RNA-binding protein determine their nucleocytoplasmic distribution in male germ cells. J Biol Chem 2004;279:31514–23.

[25] Tian Y, Simanshu DK, Ascano M, Diaz-Avalos R, Park AY, Juranek SA, et al. Multimeric assembly and biochemical characterization of the TRAX-translin endonuclease complex. Nat Struct Mol Biol 2011;18:658–64.

[26] Li L, Gu W, Liang C, Liu Q, Mello CC, Liu Y. The Translin-TRAX complex (C3PO) is a ribonuclease in tRNA processing. Nat Struct Mol Biol 2012;19:824–30.

[27] Fire A, Xu S, Montgomery MK, Kostas SA, Driver SE, Mello CC. Potent and specific genetic interference by double-stranded RNA in *Caenorhabditis elegans*. Nature 1998;391:806–11.

[28] Meister G, Tuschl T. Mechanisms of gene silencing by double-stranded RNA. Nature 2004;431:343–9.

[29] Bernstein E, Caudy AA, Hammond SM, Hannon GJ. Role for a bidentate ribonuclease in the initiation step of RNA interference. Nature 2001;409:363–6.

[30] Liu Q, Rand TA, Kalidas S, Du F, Kim HE, Smith DP, et al. R2D2, a bridge between the initiation and effector steps of the *Drosophila* RNAi pathway. Science 2003;301:1921–5.

[31] Lee YS, Nakahara K, Pham JW, Kim K, He Z, Sontheimer EJ, et al. Distinct roles for *Drosophila* Dicer-1 and Dicer-2 in the siRNA/miRNA silencing pathways. Cell 2004;117:69–81.

[32] Ding SW, Voinnet O. Antiviral immunity directed by small RNAs. Cell 2007;130: 413–26.

[33] Okamura K, Lai EC. Endogenous small interfering RNAs in animals. Nat Rev Mol Cell Biol 2008;9:673–8.

[34] Martinez J, Patkaniowska A, Urlaub H, Luhrmann R, Tuschl T. Single-stranded antisense siRNAs guide target RNA cleavage in RNAi. Cell 2002;110:563–74.

[35] Haley B, Zamore PD. Kinetic analysis of the RNAi enzyme complex. Nat Struct Mol Biol 2004;11:599–606.

[36] Liu J, Carmell MA, Rivas FV, Marsden CG, Thomson JM, Song JJ, et al. Argonaute2 is the catalytic engine of mammalian RNAi. Science 2004;305:1437–41.

[37] Song JJ, Smith SK, Hannon GJ, Joshua-Tor L. Crystal structure of Argonaute and its implications for RISC slicer activity. Science 2004;305:1434–7.

[38] Rivas FV, Tolia NH, Song JJ, Aragon JP, Liu J, Hannon GJ, et al. Purified Argonaute2 and an siRNA form recombinant human RISC. Nat Struct Mol Biol 2005;12:340–9.

[39] Pham JW, Pellino JL, Lee YS, Carthew RW, Sontheimer EJ. A Dicer-2-dependent 80s complex cleaves targeted mRNAs during RNAi in Drosophila. Cell 2004;117:83–94.

[40] Liu X, Jiang F, Kalidas S, Smith D, Liu Q. Dicer-2 and R2D2 coordinately bind siRNA to promote assembly of the siRISC complexes. RNA 2006;12:1514–20.

[41] Tomari Y, Matranga C, Haley B, Martinez N, Zamore PD. A protein sensor for siRNA asymmetry. Science 2004;306:1377–80.

[42] Miyoshi T, Takeuchi A, Siomi H, Siomi MC. A direct role for Hsp90 in pre-RISC formation in Drosophila. Nat Struct Mol Biol 2010;17:1024–6.

[43] Iwasaki S, Kobayashi M, Yoda M, Sakaguchi Y, Katsuma S, Suzuki T, et al. Hsc70/Hsp90 chaperone machinery mediates ATP-dependent RISC loading of small RNA duplexes. Mol Cell 2010;39:292–9.

[44] Iki T, Yoshikawa M, Nishikiori M, Jaudal MC, Matsumoto-Yokoyama E, Mitsuhara I, et al. In vitro assembly of plant RNA-induced silencing complexes facilitated by molecular chaperone HSP90. Mol Cell 2010;39:282–91.

[45] Maiti M, Lee HC, Liu Y. QIP, a putative exonuclease, interacts with the Neurospora Argonaute protein and facilitates conversion of duplex siRNA into single strands. Genes Dev 2007;21:590–600.

[46] Yuan YA, Yang X. High resolution crystal structure of C3PO; 2011 RCSB Protein Data Bank. Accession code: 3AXJ.

[47] Pascal JM, Hart PJ, Hecht NB, Robertus JD. Crystal structure of TB-RBP, a novel RNA-binding and regulating protein. J Mol Biol 2002;319:1049–57.

[48] Sugiura I, Sasaki C, Hasegawa T, Kohno T, Sugio S, Moriyama H, et al. Structure of human translin at 2.2 A resolution. Acta Crystallogr D Biol Crystallogr 2004;60:674–9.

[49] Gupta GD, Makde RD, Rao BJ, Kumar V. Crystal structures of Drosophila mutant translin and characterization of translin variants reveal the structural plasticity of translin proteins. FEBS J 2008;275:4235–49.

CHAPTER FOUR

Structure and Mechanism of Argonaute Proteins

Nicole T. Schirle, Ian J. MacRae[1]

Department of Molecular Biology, The Scripps Research Institute, La Jolla, California, USA
[1]Corresponding author: e-mail address: macrae@scripps.edu

Contents

Abstract

RNA silencing refers to a group of widespread gene-regulatory pathways rooted in many facets of eukaryotic biology, including brain development, stem-cell and germ-line maintenance, and cancer progression. At the molecular level, RNA-silencing pathways, such as the RNA interference (RNAi) and microRNA (miRNA) regulatory pathways, are mediated by a specialized family of RNA-binding proteins named Argonaute. Argonaute proteins are uniquely capable of binding small regulatory RNAs and using the encoded sequence information to locate and silence complementary target RNAs. The versatility and power of RNA silencing arises from the fact that Argonaute can be loaded with a small RNA of any sequence and thus can be programmed to target essentially any RNA for silencing. Here, we review current understanding of Argonaute proteins in terms of molecular structure, function, and mechanism.

1. INTRODUCTION

RNA silencing is a eukaryotic mechanism of gene silencing that functions in the cellular control of gene expression, in defense against viral infection [1], and in protection of the genome against mobile repetitive

The Enzymes, Volume 32
ISSN 1874-6047
http://dx.doi.org/10.1016/B978-0-12-404741-9.00004-0

DNA sequences [2], retro elements, and transposons [3]. All of these processes are mediated by a family of related ribonucleic-protein assemblies, generically termed RNA-induced silencing complex (RISC). The core feature of all RISCs is a member of the Argonaute protein family bound to a small guide RNA, 21–31 nucleotides in length [4,5]. Argonaute proteins use small RNAs to identify RNA targets through Watson–Crick base-pairing interactions. Depending on the Argonaute protein and biological context, mRNA targets are either cleaved by the Argonaute endonucleolytic "slicing" reaction or translationally repressed and degraded through the recruitment of mRNA degradation complexes [6]. Some Argonaute proteins function in the nucleus where they have been found to mediate cotranscriptional cleavage of pre-mRNAs [7], associate with nascent transcripts to direct chromatin and/or DNA modifications [8,9], or recruit RNA-dependent RNA polymerases to generate double-stranded RNA (dsRNA), which then feeds back into the RNAi pathway [10].

The small RNAs loaded into Argonaute are often generated from dsRNA by the RNase III enzyme Dicer [11]. Dicer cleaves long dsRNA into 21–23 nucleotide duplexes called small interfering RNAs (siRNA), and endogenous RNA hairpins into microRNAs (miRNA). Dicer products are duplexes with $5'$-phosphates and $3'$-hydroxyl functional groups. Both siRNAs and miRNAs are loaded into Argonaute as RNA duplexes. After loading, one small RNA strand is retained as the guide RNA and the other is discarded as the passenger. Guide RNAs can also be generated from single-strand RNA precursors, as is the case for piwi-interacting RNAs (piRNA) [3].

2. BASIC ARGONAUTE ARCHITECTURE

Argonaute proteins are found in all three kingdoms of life [12]. Eukaryotic Argonautes are the best studied and can be classified into three clades, termed Argonaute, Piwi, and Class III or Wago (worm-specific Argonaute), based on sequence homology [13]. Despite diversity in sequence and physiological function, all Argonaute proteins are believed to have the same overall three-dimensional architecture [14–17]. This conserved structure is composed of four globular domains and two linker domains that are organized in a bilobed architecture (Fig. 4.1). The N-terminal lobe contains the globular N and PAZ (PIWI–Argonaute–Zwille) domains connected by the L1 linker domain, while the C-terminal lobe is composed of the MID (middle) and PIWI (P-body-induced wimpy testes)

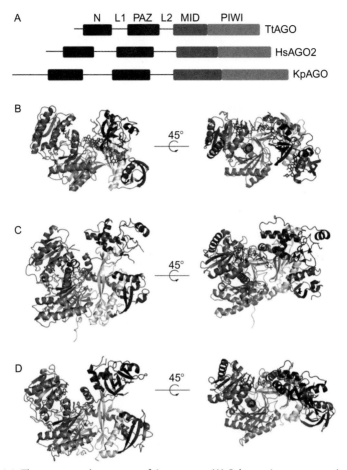

Figure 4.1 The conserved structure of Argonaute. (A) Schematic representation of the primary sequence of Argonaute proteins from three kingdoms of life. Tt Ago, *Thermus thermophilus* Argonaute; HsAgo2, human Ago2; KpAgo, budding yeast Argonaute. (B) Crystal structure of *Thermus thermophilus* Argonaute bound to a guide DNA (PDB code 3DLH). (C) Crystal structure of human Ago2 bound to a guide RNA (PDB code 4EI1). (D) Crystal structure of yeast Argonaute (PDB code 4F1N). Guide DNA or RNA molecules are colored red. (See color plate section in the back of the book.)

domains [12,18]. The two lobes are connected by the L2 linker, which forms the base of the central cleft. The two lobes create a central cleft that harbors the slicer active site and cradles guide and target nucleic acid molecules.

The major structural differences between prokaryotic and eukaryotic Argonaute proteins that likely influence the binding of guide and target molecules are seen in the orientations of the globular domains within each lobe.

The N and PAZ domains of eukaryotic Argonautes are rotated about the L2 domain relative to the corresponding domains of prokaryotes [14,19]. These differences give eukaryotic Argonautes a wider central cleft than the cleft of prokaryotic Argonaute. It has been noted that the wider central cleft found in eukaryotic Argonaute proteins may be "open ended," which could allow the binding of guide/target duplexes that extend outside the protein and into the bulk solvent [19]. Prokaryotic Argonautes have been observed in two distinct conformations, where the central cleft is in a more closed conformation upon target binding and widens to an open conformation upon target recognition [15,20].

All domains of Argonaute, including the linker domains, contact the guide RNA, while defined binding pockets in the MID and PAZ domains recognize the guide RNA termini. The 5′-phosphate of the guide RNA is buried in a hydrophilic pocket within the MID domain and the identity of the 5′-base is probed by a rigid loop termed the nucleotide specificity loop. The 3′-end of the guide RNA binds Argonaute proteins in a shallow pocket formed by a solvent-exposed portion of the PAZ domain. The primary interactions with the 3′-end are made with the sugar ring, and therefore provide little or no selection of 3′-base identity. The nucleotides 2–7 of the guide RNA (from the 5′-end) are termed the "seed region" and are the most important part of the guide for target recognition. The seed region of the guide is preorganized in an A-form helical conformation by Argonaute through an extensive array of salt bridges and hydrogen bonds to the phosphate backbone. Nucleotides 2–6 of the seed are exposed to the solvent in order to facilitate Watson–Crick base pairing to target RNAs. The rest of the guide RNA is threaded through the center of the protein, where it is cradled between the two lobes and makes extensive hydrogen bond contacts to the hydrophilic channel of the protein.

3. RECOGNITION OF THE GUIDE 5′-END BY THE MID DOMAIN

A 5′-phosphate has long been recognized as an important feature of siRNAs and has been suggested to function as a licensing factor for entry of small RNAs into the RNAi pathway [21]. Crystal structures of full-length Argonaute proteins from both prokaryotes and eukaryotes revealed that the 5′-phosphate of the guide is buried within a conserved pocket in the MID domain [14–16,19,20,22–24]. Conserved tyrosine, lysine, and glutamine residues from both the MID and PIWI domains, as well as the C-terminal carboxyl group, form an extensive network of salt linkages and hydrogen bonds with the 5′-phosphate. These interactions position the 5′-base of the guide out of the helical stack observed in the following nucleotides of the guide and into the MID domain where it stacks against a well-conserved tyrosine residue. Therefore, the first 5′-base of guide RNAs is not available for pairing to target RNAs, which explains why the first base in miRNAs does not significantly contribute to target recognition [25].

Sequencing of Argonaute-associated small RNAs revealed that Argonaute proteins often display 5′-nucleotide identity preferences for binding guide RNAs [26–29]. For example, human Ago2 usually associates with miRNAs having either a 5′-uridine (U) or adenosine (A), while Arabidopsis Argonaute proteins 1, 2, 4, and 5 preferentially associate with guide RNAs bearing 5′ U, A, A, and cytosine (C), respectively [30–32]. The level of 5′-nucleotide selectivity varies between Argonaute proteins, and in some cases, the bias for a particular nucleotide can be very high. For example, >98% of the endogenous small RNAs associated with *Schizosaccharomyces pombe* Ago1 were found to contain a 5′ U [33]. In contrast, *Arabidopsis* Ago7 does not have any clear 5′-nucleotide preference [32,34]. The observed 5′-nucleotide preferences correlate with differences in the affinities of the MID domains of Argonaute proteins for various possible nucleotides. These affinities are thought to contribute to guide strand selection during the process of RISC loading [26–28]. Supporting this idea, biophysical analysis using nuclear magnetic resonance revealed that the MID domain of Ago2 binds A and U with an affinity about 20 times greater than G or C [35].

The X-ray crystal structures of the human Ago2 MID domain in complex with nucleotide monophosphates revealed a Rossman-fold that recognizes the 5′-monophosphate of guide RNAs and the base edges of A and U (Fig. 4.2) [35]. The Watson–Crick faces of AMP and UMP are recognized by a rigid loop termed the nucleotide specificity loop. The hydrogen bond acceptors of the nucleotide rings (O4 and N1 atoms from AMP and UMP, respectively) coincide in space, allowing either to make hydrogen bonds with the backbone amide group of Thr-526 the human Ago2 MID domain. In contrast, base edges of GMP and CMP sterically clash with the specificity loop. Similarly, structural studies of the MID domains from *Arabidopsis* Argonaute proteins showed that in *Arabidopsis* Ago1, the Asn-687 of the specificity loop recognizes O_2 of U and C nucleotides [32]. In a clever experiment, Zha and coworkers demonstrated that mutating the specificity loop of the *Arabidopsis* Ago1 MID domain, making the Ago1 specificity loop resemble the loop in *Arabidopsis* Ago4, can change the binding preference from U to A [36].

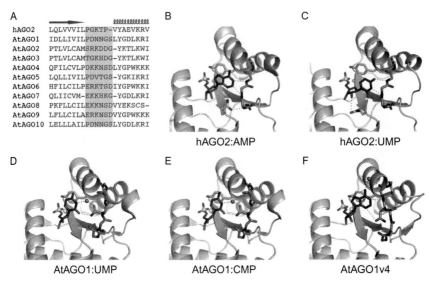

A
hAGO2	LQLVVVILPGKTP-VYAEVKRV
AtAGO1	IDLLIVILPDNNGSLYGDLKRI
AtAGO2	PTLVLCAMSRKDDG-YKTLKWI
AtAGO3	PTLVLCAMTGKHDG-YKTLKWI
AtAGO4	QFILCVLPDKKNSDLYGPWKKK
AtAGO5	LQLLIVILPDVTGS-YGKIKRI
AtAGO6	HFILCILPERKTSDIYGPWKKI
AtAGO7	QLIICVM-EKKHKG-YGDLKRI
AtAGO8	PKFLLCILEKKNSDVYEKSCS-
AtAGO9	LFLLCILAERKNSDVYGPWKKK
AtAGO10	LELLLAILPDNNGSLYGDLKRI

B

C

hAGO2:AMP   hAGO2:UMP

D   E   F

AtAGO1:UMP   AtAGO1:CMP   AtAGO1v4

Figure 4.2 Recognition of the guide 5′-nucleotide. (A) Sequence alignment of the MID domain selectivity loop from human Ago2 (hAgo2) and the 10 *Arabidopsis thaliana* (At) Argonautes. (B) and (C) Human Ago2 MID domain bound to AMP and UMP, respectively (PDB codes 3LUD and 3LUJ). (D) and (E) AtAGO1 MID domain bound to UMP and CMP, respectively (PDB codes 4G0P and 4G0Q). (F) Crystal structure of mutated AtAGO1 (AtAGO1v4) with AMP modeled (MID domain PDB code 3VNB). Selectively loop residues are shown in stick representation. Hydrogen bonds are indicated by dashed lines and water molecules are shown as pink spheres. (See color plate section in the back of the book.)

The general theme is that Argonaute proteins recognize the identity of the 5′-nucleotide of guide RNAs through direct interactions with the specificity loop in their MID domains. Subtle differences in specificity loop structure can give rise to the recognition and loading of guide RNAs with different 5′-nucleotide identities. These differences, in combination with small RNA duplex structure, can be used by cells to target specific small RNAs to specific Argonaute proteins [28,30].

4. PREORGANIZATION OF THE SEED FACILITATES TARGET RECOGNITION

A fundamental property of Argonaute proteins that is critical to their function is the ability to specifically recognize target RNAs that bear sequence complementary to their loaded guides. *In vitro* kinetic analyses of RISC activity have shown that Argonaute proteins carry out this task with remarkable efficiency [37,38]. Indeed, one study found that human RISC can bind and cleave a target RNA 10 times faster than a naked guide RNA anneals to a complementary target RNA in free solution [38]. The ability of Argonaute to efficiently identify and bind to targets has been largely attributed to the idea that Argonaute prearranges the seed region (nucleotides 2–7) of the guide RNA in an A-form configuration [25,39,40]. Preorganization of the seed region is thought to allow the protein to efficiently scan potential target RNAs for complementary sequences while reducing the entropic cost associated with forming a stable duplex with target RNAs [41]. The reduced entropy associated with pairing to a preordered seed region has also been attributed to the importance of guide nucleotides 2–7 for recognition of microRNA target sites [25].

Structural studies of Argonaute proteins support the "seed-pairing" model of target recognition. In crystal structures of both human Ago2 and budding yeast Ago1, nucleotides 2–6 of bound guide RNAs were prearranged in an A-form helical conformation with the solvent-exposed Watson–Crick edges positioned for nucleation with the RNA targets (Fig. 4.3) [14,19]. Bases 2–6 of guide RNAs are organized through extensive salt linkages and hydrogen bonds between the phosphate backbone and the PIWI domain of Argonaute. Importantly, except for the 5′-nucleotide, Argonaute does not make any direct interactions with the guide RNA bases. This enables guide RNAs of any nucleotide sequence to be used by Argonaute. The ability of Argonaute to bind and use guide RNAs of any sequence means that RISC can be programmed to silence essentially any unique nucleotide target sequence. Thus, the structural architecture of Argonaute, which allows the protein to tightly associate with small RNAs in a sequence-independent manner, makes the RNA-silencing pathways versatile and useful in many facets of biology.

Structures of eukaryotic Argonaute proteins from yeast and human also revealed that an alpha helix (termed helix 7) in the L2 linker domain introduces a kink in the guide RNA between bases 6 and 7 that disrupts helical stacking between these two nucleotides (Fig. 4.3). The kink is introduced by a well-conserved isoleucine residue, which is inserted between the faces of bases 6 and 7. Structures of Argonaute bound to a guide and target RNA are not yet available, and while its function is not yet established, the kink between bases 6 and 7 has intriguing mechanistic implications. One possibility is that the crystallized structures represent product release states of Argonaute. In this model, Argonaute introduces the kink to stabilize a conformation in the guide RNA that is less prone to form a duplex with target RNAs or cleavage products. Consistent with this idea, modeling an A-form duplex onto the guide RNA suggests that helix-7 would have to shift away

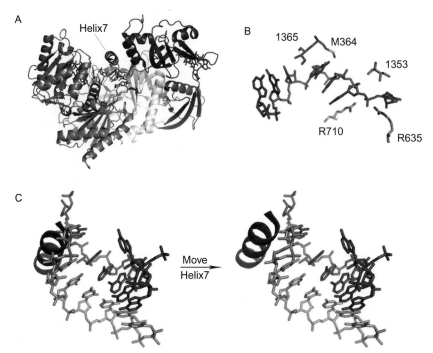

Figure 4.3 Guide RNA binding by human Ago2. (A) Crystal structure of human Ago2 bound to guide RNA, miR-20a (PDB code 4F3T). Domains are colored as in Fig. 4.1. (B) Distinct kinks in the guide RNA (red) are stabilized by direct interactions with the protein. (C) Target RNA (violet) modeled onto the guide RNA clashes with helix7 of Ago2 (PDB code 4EI1). Movement of helix7 through conformational change (modeled here) may be a feature of target RNA recognition. (See color plate section in the back of the book.)

from the guide to sterically accommodate target binding, thereby releasing the constraints placed on nucleotide 7 (Fig. 4.3C). Target cleavage via the slicing reaction could introduce conformational flexibility and allow Ago2 to reintroduce the kink, thereby destabilizing the duplex between guide and product RNAs and facilitating product release. The kink may also facilitate miRNA bulged-site pairing, a recently proposed alternative mode of miRNA-target interaction [42].

5. RECOGNITION OF THE GUIDE RNA 3′-END BY THE PAZ DOMAIN

Argonaute also binds to the 3′-end of guide RNAs in a sequence-independent manner. 3′-end recognition is mediated by the PAZ domain, a small RNA-binding domain found in Argonaute and Dicer proteins [12].

PAZ is composed of a small β-barrel that is a distant member of the oligonucleotide/oligosaccharide-binding fold. PAZ domains specifically recognize dsRNA ends containing a 3′-two-base overhang [43–45]. Argonaute PAZ domains contain a binding pocket that interacts with the 3′-end of guide RNAs through π-stacking with the terminal base, hydrogen bonding to the 2′- and 3′-hydroxyls, and Van der Waals interactions with the sugar [44–46]. The PAZ domain also forms hydrogen bonds and salt linkages to the sugar–phosphate backbone of two nucleotides preceding the terminal nucleotide. As with all other protein–guide RNA interactions in Argonaute (except for recognition of the 5′-base by the specificity loop), no sequence-specific contacts are made to the guide RNA by the PAZ domain.

6. THE TWO-STATE MODEL OF TARGET BINDING

In contrast to MID domain binding of the 5′-end of the guide RNA, 3′-end binding to PAZ is believed to be a dynamic process, with rounds of binding and release as Argonaute engages target RNAs. The crystal structure of human Ago2 bound to miR-20a shows the 3′-end of the guide miRNA bound by the PAZ domain when Argonaute is not engaged with a target RNA [47]. However, if both the 5′- and 3′-ends of the guide RNA remain securely bound to the protein, it would be topologically impossible to form a full 20-base pair duplex structure between guide- and target-RNAs. A two-state model for target binding has therefore been proposed [16,48,49]. In this model, recognition of a target RNA by the 5′ half of the guide RNA prompts the release of the 3′-end of the guide RNA from the PAZ domain. The 3′-end is then topologically free to rotate around the engaged target, allowing the formation of a duplex structure (Fig. 4.4). In cases in which only the seed region is complementary to the target, it is believed that the 3′-end of the guide may stay associated with the PAZ domain [15,50].

The best structural evidence for the two-state model of target binding by Argonaute comes from studies of the Argonaute protein from the bacteria *Thermus thermophilus*. Unlike its eukaryotic cousins, the *Thermus thermophilus* Argonaute uses a guide DNA instead of RNA and can cleave both RNA and DNA targets. In a series of crystal structures containing a guide DNA and DNA or RNA target molecules, Wang and coworkers were able to generate high-resolution "snap-shots" of *Thermus thermophilus* binding to a target (Fig. 4.4) [15,20,23]. These structures indicated that, as proposed, the initial pairing of guide and target occurs at the 5′-end of the guide in the

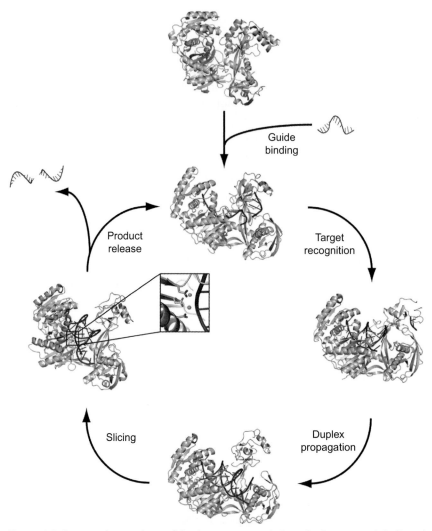

Figure 4.4 Structural snap-shots of the Argonaute catalytic cycle. Argonaute is believed to pass through distinct conformations during catalysis. These include the empty enzyme (PDB code 1U04) binding to a guide (PDB code 3DLH), target recognition by the seed region (PDB code 3HO1), duplex propagation accompanied by release of the guide 3′-end (PDB code 3HK2), target cleavage (PDB code 3HVR), and product release. (See color plate section in the back of the book.)

seed region. The duplex then extends down the central cleft of the protein, ultimately pulling the 3′-end of the guide out of the PAZ domain.

Interestingly, structural studies of *Thermus thermophilus* Argonaute revealed that a fully paired, 15-base pair guide–target duplex fills the entire central cleft of the protein [23]. This observation suggests that the protein cannot accommodate a duplex structure between guide and target beyond nucleotide 16 of the guide DNA. That is, *Thermus thermophilus* Argonaute appears to only be large enough to bind to duplexes 15 base pairs in length. In this structure, the 3′-end of the guide DNA abuts the N domain of Argonaute, which appears to sterically block any further duplex formation between guide and target (Fig. 4.4). It is not yet clear if similar constraints exist in eukaryotic Argonaute proteins. However, it has been noted that the cleft between the N and PIWI domains of eukaryotic Argonaute is larger than that seen in bacteria [19]. It is therefore possible that eukaryotic Argonaute proteins can accommodate longer guide:target duplexes than seen in bacteria.

7. THE RUBBER BAND MODEL OF RISC LOADING

Hydrolysis of ATP has long been known to be important for establishing RISC activity in cell extracts [21,51]. Several steps in RISC formation have been shown to utilize ATP, including small RNA production [21,52,53], small RNA phosphorylation (in the case of synthetic small RNAs) [54], and duplex loading into Argonaute and passenger strand removal [21,55]. The ATP requirement in the final steps of RISC maturation is likely due, at least in part, to the function of Hsc70/Hsp90 chaperone machinery in loading small RNA duplexes into Argonaute [56,57]. Importantly, the loading of small RNA duplexes into Argonaute requires ATP, but separating the two small RNA strands within Argonaute does not [57]. The combined observations have led to the "rubber band" model for RISC loading, in which chaperone proteins use energy from ATP hydrolysis to pull empty Argonaute proteins into a conformationally "stretched" or strained state. The release from this state is thought to drive unwinding of the guide–passenger duplex during RISC loading [57,58].

Consistent with the rubber band model, the structure of *Thermus thermophilus* Argonaute bound to a guide DNA and fully complementary target suggests that the bacterial protein can only accommodate guide–target duplexes about 15 base pairs in length [23]. In this structure, the end of the guide–target duplex abuts the N domain of Argonaute, with the base of guide nucleotide 16 stacked against tyrosine-43 in the N domain, and the paired base in the target RNA stacked against proline-44. Based on this observation, it was proposed that the N domain might be involved in modulating guide–target pairing at the $3'$-end of the guide and be important for unwinding the guide–passenger RNA duplex during RISC loading [58]. In this model, the N domain of Argonaute functions as a wedge that drives the two strands apart. This model is supported by the finding that mutations disrupting the structure of the N domain of human Ago2 inhibit small RNA duplex unwinding and the formation of active RISC [59].

8. THE SLICER ACTIVE SITE RESIDES IN THE PIWI DOMAIN

Pioneering structural studies of the Argonaute protein from *Pyrococcus furiosus* revealed that the PIWI domain contains an RNase H-like fold [17]. This observation led to the discovery that Argonaute is the endoribonuclease "Slicer" of the RNAi pathway and that the slicer active site resides in the PIWI domain [16,17,60]. Like RNase H [61], the slicer active site contains four catalytic residues that coordinate two divalent cations (Fig. 4.4) [19]. Therefore, like many other nucleases [62–64], Argonaute likely employs a two-metal mechanism for RNA hydrolysis [61].

Currently, the crystal structure of *Thermus thermophilus* Argonaute guide–target duplex is the only available high-resolution characterization of a scissile phosphate in a slicer active site [23]. By crystallizing Argonaute bound to a guide DNA in complex with a fully complementary target RNA, Wang and coworkers were able to visualize the phosphodiester backbone of target nucleotides $10'$ and $11'$ (the target bases paired with DNA guide nucleotides 10 and 11) in the slicer active site. Under low magnesium conditions, a single magnesium ion, proposed to facilitate nucleophilic attack, was observed coordinated to two catalytic aspartates and the scissile phosphate. Increasing the concentration of magnesium in the crystallization conditions promoted binding of a second magnesium ion that was coordinated to two active site aspartates and the scissile phosphate and was proposed to facilitate leaving of the cleaved phosphate. The structure of the *Thermus thermophilus* Argonaute active site was shown to superimpose well with the *Bacillus halodurans* RNase H active site, suggesting that observed magnesium ions support the same mechanism of two-metal-ion hydrolysis of the phosphodiester backbone in the two enzyme families.

Comparison of the PIWI domain structures of Ago1 from the budding yeast *Kluyveromyces polysporus* and QDE-2 (an Argonaute protein from the filamentous fungus *Neurospora crassa*) revealed that the fourth catalytic residue (glutamate-1013 in *Kluyveromyces polysporus*) resides on a flexible loop that can move into and out of the active site [19,65]. The flexible loop has been termed a "glutamate finger," which has the ability to "plug-in" to the active tetrad, forming an extensive network of hydrogen bonds, which hold the loop in a rigid conformation. Supporting this model, all of the conformations of *Thermus thermophilus* Argonaute that are believed to be active (with the 3'-end of the guide DNA released from the PAZ domain) have the glutamate finger "plugged-in," whereas crystallized conformations that appear to be inactive have an "unplugged" or "plugged-out" glutamate finger. This, combined with yeast knock-in biochemical evidence, suggests that the glutamate finger functions as the fourth catalytic residue of the catalytic tetrad [19]. Analysis of the human Ago2 crystal structure also reveals a "plugged-in" glutamate finger, suggesting that the active site tetrad is conserved in mammals as well [14]. In human Ago2, the catalytic tetrad is composed of residues D597, E637, D669, and H807.

Conservation of a pluggable glutamate finger from bacteria to eukaryotes indicates that the mobility of this element may be important to Argonaute function. It was proposed that the unplugged conformation is involved in the loading of apo-Argonaute proteins with siRNA duplexes [19]. In this model, release of the guide 3'-end from the PAZ domain induces a switch to the plugged-in conformation and subsequent cleavage and removal of the passenger strand. It is also possible that the unplugged conformation is involved in the fidelity of target recognition by contributing to the avoidance of slicing noncognate target RNAs.

9. RECRUITMENT OF SILENCING FACTORS

Although the slicing reaction is sufficient to silence an mRNA target, Argonaute proteins often function by recruiting additional protein factors to targeted RNA transcripts. The ability of Argonaute to associate with diverse cellular factors has allowed divergent RNA-silencing mechanisms to evolve and integrate into many facets of eukaryotic biology [66]. Indeed, various Argonaute proteins have been found to associate with many different protein factors, including RNA helicases [55,67–69], nucleases [55,70–72], RNA polymerases [10,73], chromodomain proteins and methyltransferases [8,67,73], and a variety of RNA-binding proteins [72,74].

Many Argonaute–protein interactions are sensitive to RNase treatment, suggesting these may be indirect [74]. However, there is also clear evidence for some direct Argonaute–protein interactions [8,74]. The best-characterized Argonaute-binding proteins contain unstructured glycine-tryptophan (GW)-rich regions [8,75–77]. In flies, Ago1 recruits GW182, a GW-rich protein whose interaction with Argonaute is mediated by tryptophan [75,78,79]. Specifically, the N-terminal domain of GW182 interacts with the PIWI domain of the *Drosophila melanogaster* Ago1 during miRNA-mediated repression, while the C-terminal region of GW182 is required for repression and localization of Ago1 and GW182 to P-bodies [78,79]. GW182 recruits mRNA-decay factors, such as the DCP1 and DCP2 decapping enzymes and the PAN2–PAN3 and CCR4: NOT deadenylase complexes, to targeted mRNAs [80–83]. In humans, the GW182 paralogs are the trinucleotide repeat-containing genes (TNRC6A-C). Similar to fly GW182, TNRC6 proteins have been shown to interact with the human Ago PIWI domain through tryptophan-mediated interactions [75].

The crystal structure of human Ago2 bound to tryptophan revealed a likely interaction surface for TNRC6 binding [14]. Two tryptophan molecules were observed bound in adjacent pockets of the PIWI domain. The tryptophans were both bound such that the indole ring side-chains were buried within the protein and the peptide backbone atoms were solvent exposed on the surface of the protein (Fig. 4.5). It is interesting to note that many of the tryptophans in the N-terminal Argonaute interaction region of GW182 and TNRC6B occur in pairs separated by 8–14 amino acids. An extended 8-amino acid peptide spans a distance of about 24 Å, which closely matches the distance between the two tryptophan-binding sites,

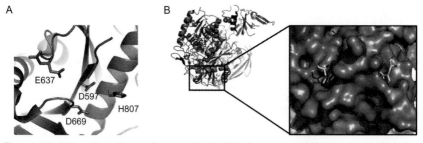

Figure 4.5 Interaction site on human Ago2. (A) Close view of the Ago2 slicer active site. Residues of the catalytic tetrad are shown in stick representation (PDB code 4EI1). (B) Tandem tryptophan-binding sites in the PIWI domain (PDB code 4EI3). A surface representation of the two sites is shown with bound tryptophan molecules as yellow sticks (right). The position of the tryptophan-binding sites relative to the entire protein is indicated on the left. (See color plate section in the back of the book.)

when measuring along the surface of Ago2. Based on these observations, it has been suggested that Argonaute may specifically recognize GW proteins by binding to tandem tryptophan residues that are separated by an appropriately sized flexible linker [14]. Although this is currently the most detailed model for an Argonaute–protein interaction, it remains to be rigorously tested. A major challenge for the future will be to determine the structural basis for recruitment of the many known Argonaute-associated proteins to targeted RNAs to obtain a general understanding of the quaternary structures of the numerous forms of RISC.

ACKNOWLEDGMENTS
We are grateful to Jessica Gruttadauria for critically reading the chapter. N. T. S. is a predoctoral fellow of the American Heart Association. I. J. M. is supported by NIH grant R01GM086701 and is a Pew Scholar in the Biomedical Sciences.

REFERENCES
[1] Baulcombe DC, Molnar A. Crystal structure of p19–a universal suppressor of RNA silencing. Trends Biochem Sci 2004;29:279.
[2] Buhler M, Moazed D. Transcription and RNAi in heterochromatic gene silencing. Nat Struct Mol Biol 2007;14:1041.
[3] Aravin AA, Hannon GJ, Brennecke J. The Piwi-piRNA pathway provides an adaptive defense in the transposon arms race. Science 2007;318:761.
[4] Hammond SM, Boettcher S, Caudy AA, Kobayashi R, Hannon GJ. Argonaute2, a link between genetic and biochemical analyses of RNAi. Science 2001;293:1146.
[5] Rivas FV, et al. Purified Argonaute2 and an siRNA form recombinant human RISC. Nat Struct Mol Biol 2005;12:340.
[6] Huntzinger E, Izaurralde E. Gene silencing by microRNAs: contributions of translational repression and mRNA decay. Nat Rev Genet 2011;12:99.
[7] Guang S, et al. Small regulatory RNAs inhibit RNA polymerase II during the elongation phase of transcription. Nature 2010;465:1097.
[8] Verdel A, et al. RNAi-mediated targeting of heterochromatin by the RITS complex. Science 2004;303:672.
[9] Zilberman D, Cao X, Jacobsen SE. ARGONAUTE4 control of locus-specific siRNA accumulation and DNA and histone methylation. Science 2003;299:716.
[10] Motamedi MR, et al. Two RNAi complexes, RITS and RDRC, physically interact and localize to noncoding centromeric RNAs. Cell 2004;119:789.
[11] Bernstein E, Caudy AA, Hammond SM, Hannon GJ. Role for a bidentate ribonuclease in the initiation step of RNA interference. Nature 2001;409:363.
[12] Cerutti L, Mian N, Bateman A. Domains in gene silencing and cell differentiation proteins: the novel PAZ domain and redefinition of the Piwi domain. Trends Biochem Sci 2000;25:481.
[13] Joshua-Tor L. The Argonautes. Cold Spring Harb Symp Quant Biol 2006;71:67.
[14] Schirle NT, MacRae IJ. The crystal structure of human Argonaute2. Science 2012;336:1037.
[15] Wang Y, et al. Structure of an argonaute silencing complex with a seed-containing guide DNA and target RNA duplex. Nature 2008;456:921.

[16] Yuan YR, et al. Crystal structure of A. aeolicus argonaute, a site-specific DNA-guided endoribonuclease, provides insights into RISC-mediated mRNA cleavage. Mol Cell 2005;19:405.

[17] Song JJ, Smith SK, Hannon GJ, Joshua-Tor L. Crystal structure of Argonaute and its implications for RISC slicer activity. Science 2004;305:1434.

[18] Lin H, Spradling AC. A novel group of pumilio mutations affects the asymmetric division of germline stem cells in the Drosophila ovary. Development 1997;124:2463.

[19] Nakanishi K, Weinberg DE, Bartel DP, Patel DJ. Structure of yeast Argonaute with guide RNA. Nature 2012;486:368.

[20] Wang Y, Sheng G, Juranek S, Tuschl T, Patel DJ. Structure of the guide-strand-containing argonaute silencing complex. Nature 2008;456:209.

[21] Nykanen A, Haley B, Zamore PD. ATP requirements and small interfering RNA structure in the RNA interference pathway. Cell 2001;107:309.

[22] Ma JB, et al. Structural basis for 5'-end-specific recognition of guide RNA by the A. fulgidus Piwi protein. Nature 2005;434:666.

[23] Wang Y, et al. Nucleation, propagation and cleavage of target RNAs in Ago silencing complexes. Nature 2009;461:754.

[24] Parker JS, Roe SM, Barford D. Structural insights into mRNA recognition from a PIWI domain-siRNA guide complex. Nature 2005;434:663.

[25] Lewis BP, Burge CB, Bartel DP. Conserved seed pairing, often flanked by adenosines, indicates that thousands of human genes are microRNA targets. Cell 2005;120:15.

[26] Ghildiyal M, et al. Endogenous siRNAs derived from transposons and mRNAs in Drosophila somatic cells. Science 2008;320:1077.

[27] Seitz H, Ghildiyal M, Zamore PD. Argonaute loading improves the 5' precision of both MicroRNAs and their miRNA★ strands in flies. Curr Biol 2008;18:147.

[28] Ghildiyal M, Xu J, Seitz H, Weng Z, Zamore PD. Sorting of Drosophila small silencing RNAs partitions microRNA★ strands into the RNA interference pathway. RNA 2010;16:43.

[29] Lau NC, Lim LP, Weinstein EG, Bartel DP. An abundant class of tiny RNAs with probable regulatory roles in Caenorhabditis elegans. Science 2001;294:858.

[30] Mi S, et al. Sorting of small RNAs into Arabidopsis argonaute complexes is directed by the 5' terminal nucleotide. Cell 2008;133:116.

[31] Takeda A, Iwasaki S, Watanabe T, Utsumi M, Watanabe Y. The mechanism selecting the guide strand from small RNA duplexes is different among argonaute proteins. Plant Cell Physiol 2008;49:493.

[32] Frank F, Hauver J, Sonenberg N, Nagar B. Arabidopsis Argonaute MID domains use their nucleotide specificity loop to sort small RNAs. EMBO J 2012;31:3588.

[33] Buhler M, Spies N, Bartel DP, Moazed D. TRAMP-mediated RNA surveillance prevents spurious entry of RNAs into the Schizosaccharomyces pombe siRNA pathway. Nat Struct Mol Biol 2008;15:1015.

[34] Montgomery TA, et al. Specificity of ARGONAUTE7-miR390 interaction and dual functionality in TAS3 trans-acting siRNA formation. Cell 2008;133:128.

[35] Frank F, Sonenberg N, Nagar B. Structural basis for 5'-nucleotide base-specific recognition of guide RNA by human AGO2. Nature 2010;465:818.

[36] Zha X, Xia Q, Adam Yuan Y. Structural insights into small RNA sorting and mRNA target binding by Arabidopsis Argonaute Mid domains. FEBS Lett 2012;586:3200.

[37] Haley B, Zamore PD. Kinetic analysis of the RNAi enzyme complex. Nat Struct Mol Biol 2004;11:599.

[38] Ameres SL, Martinez J, Schroeder R. Molecular basis for target RNA recognition and cleavage by human RISC. Cell 2007;130:101.

[39] Bartel DP. MicroRNAs: genomics, biogenesis, mechanism, and function. Cell 2004;116:281.

[40] Lambert NJ, Gu SG, Zahler AM. The conformation of microRNA seed regions in na-tive microRNPs is prearranged for presentation to mRNA targets. Nucleic Acids Res 2011;39:4827.

[41] Parker JS, Parizotto EA, Wang M, Roe SM, Barford D. Enhancement of the seed-target recognition step in RNA silencing by a PIWI/MID domain protein. Mol Cell 2009;33:204.

[42] Chi SW, Hannon GJ, Darnell RB. An alternative mode of microRNA target recogni-tion. Nat Struct Mol Biol 2012;19:321.

[43] Ma JB, Ye K, Patel DJ. Structural basis for overhang-specific small interfering RNA recognition by the PAZ domain. Nature 2004;429:318.

[44] Song JJ, et al. The crystal structure of the Argonaute2 PAZ domain reveals an RNA binding motif in RNAi effector complexes. Nat Struct Biol 2003;10:1026.

[45] Lingel A, Simon B, Izaurralde E, Sattler M. Structure and nucleic-acid binding of the *Drosophila* Argonaute 2 PAZ domain. Nature 2003;426:465.

[46] Yan KS, et al. Structure and conserved RNA binding of the PAZ domain. Nature 2003;426:468.

[47] Elkayam E, et al. The structure of human argonaute-2 in complex with miR-20a. Cell 2012;150:100.

[48] Tomari Y, Zamore PD. Perspective: machines for RNAi. Genes Dev 2005;19:517.

[49] Filipowicz W. RNAi: the nuts and bolts of the RISC machine. Cell 2005;122:17.

[50] Ameres SL, et al. Target RNA-directed trimming and tailing of small silencing RNAs. Science 2010;328:1534.

[51] Zamore PD, Tuschl T, Sharp PA, Bartel DP. RNAi: double-stranded RNA directs the ATP-dependent cleavage of mRNA at 21 to 23 nucleotide intervals. Cell 2000;101:25.

[52] Cenik ES, et al. Phosphate and R2D2 restrict the substrate specificity of Dicer-2, an ATP-driven ribonuclease. Mol Cell 2011;42:172.

[53] Welker NC, et al. Dicer's helicase domain discriminates dsRNA termini to promote an altered reaction mode. Mol Cell 2011;41:589.

[54] Weitzer S, Martinez J. The human RNA kinase hClp1 is active on 3′ transfer RNA exons and short interfering RNAs. Nature 2007;447:222.

[55] Tomari Y, et al. RISC assembly defects in the *Drosophila* RNAi mutant armitage. Cell 2004;116:831.

[56] Iki T, et al. In vitro assembly of plant RNA-induced silencing complexes facilitated by molecular chaperone HSP90. Mol Cell 2010;39:282.

[57] Iwasaki S, et al. Hsc70/Hsp90 chaperone machinery mediates ATP-dependent RISC loading of small RNA duplexes. Mol Cell 2010;39:292.

[58] Kawamata T, Tomari Y. Making RISC. Trends Biochem Sci 2010;35:368.

[59] Kwak PB, Tomari Y. The N domain of Argonaute drives duplex unwinding during RISC assembly. Nat Struct Mol Biol 2012;19:145.

[60] Liu J, et al. Argonaute2 is the catalytic engine of mammalian RNAi. Science 2004;305:1437.

[61] Nowotny M, Gaidamakov SA, Crouch RJ, Yang W. Crystal structures of RNase H bound to an RNA/DNA hybrid: substrate specificity and metal-dependent catalysis. Cell 2005;121:1005.

[62] Macrae IJ, et al. Structural basis for double-stranded RNA processing by Dicer. Science 2006;311:195.

[63] Bujacz G, et al. Binding of different divalent cations to the active site of avian sarcoma virus integrase and their effects on enzymatic activity. J Biol Chem 1997;272:18161.

[64] Vipond IB, Baldwin GS, Halford SE. Divalent metal ions at the active sites of the EcoRV and EcoRI restriction endonucleases. Biochemistry 1995;34:697.

[65] Boland A, Huntzinger E, Schmidt S, Izaurralde E, Weichenrieder O. Crystal structure of the MID-PIWI lobe of a eukaryotic Argonaute protein. Proc Natl Acad Sci USA 2011;108:10466.

[66] Pratt AJ, MacRae IJ. The RNA-induced silencing complex: a versatile gene-silencing machine. J Biol Chem 2009;284:17897.

[67] Meister G, et al. Identification of novel argonaute-associated proteins. Curr Biol 2005;15:2149.

[68] Mourelatos Z, et al. miRNPs: a novel class of ribonucleoproteins containing numerous microRNAs. Genes Dev 2002;16(720).

[69] Robb GB, Rana TM. RNA helicase A interacts with RISC in human cells and functions in RISC loading. Mol Cell 2007;26:523.

[70] Sasaki T, Shiohama A, Minoshima S, Shimizu N. Identification of eight members of the Argonaute family in the human genome small star, filled. Genomics 2003;82:323.

[71] Caudy AA, et al. A micrococcal nuclease homologue in RNAi effector complexes. Nature 2003;425:411.

[72] Chendrimada TP, et al. TRBP recruits the Dicer complex to Ago2 for microRNA processing and gene silencing. Nature 2005;436:740.

[73] Bies-Etheve N, et al. RNA-directed DNA methylation requires an AGO4-interacting member of the SPT5 elongation factor family. EMBO Rep 2009;10:649.

[74] Hock J, et al. Proteomic and functional analysis of Argonaute-containing mRNA-protein complexes in human cells. EMBO Rep 2007;8:1052.

[75] Till S, et al. A conserved motif in Argonaute-interacting proteins mediates functional interactions through the Argonaute PIWI domain. Nat Struct Mol Biol 2007;14:897.

[76] Baillat D, Shiekhattar R. Functional dissection of the human TNRC6 (GW182-related) family of proteins. Mol Cell Biol 2009;29:4144.

[77] El-Shami M, et al. Reiterated WG/GW motifs form functionally and evolutionarily conserved ARGONAUTE-binding platforms in RNAi-related components. Genes Dev 2007;21:2539.

[78] Eulalio A, Huntzinger E, Izaurralde E. GW182 interaction with Argonaute is essential for miRNA-mediated translational repression and mRNA decay. Nat Struct Mol Biol 2008;15:346.

[79] Eulalio A, Helms S, Fritzsch C, Fauser M, Izaurralde E. A C-terminal silencing domain in GW182 is essential for miRNA function. RNA 2009;15:1067.

[80] Rehwinkel J, Behm-Ansmant I, Gatfield D, Izaurralde E. A crucial role for GW182 and the DCP1:DCP2 decapping complex in miRNA-mediated gene silencing. RNA 2005;11:1640.

[81] Behm-Ansmant I, et al. mRNA degradation by miRNAs and GW182 requires both CCR4:NOT deadenylase and DCP1:DCP2 decapping complexes. Genes Dev 2006;20:1885.

[82] Fabian MR, et al. miRNA-mediated deadenylation is orchestrated by GW182 through two conserved motifs that interact with CCR4-NOT. Nat Struct Mol Biol 2011;18:1211.

[83] Braun JE, Huntzinger E, Fauser M, Izaurralde E. GW182 proteins directly recruit cytoplasmic deadenylase complexes to miRNA targets. Mol Cell 2011;44:120.

CHAPTER FIVE

Drosha and DGCR8 in MicroRNA Biogenesis

Feng Guo[*,1]

*Department of Biological Chemistry, and Molecular Biology Institute, University of California, Los Angeles, California, USA
[1]Corresponding author: e-mail address: fguo@mbi.ucla.edu

Contents

Abstract

Drosha and DGCR8 (also called Pasha), together known as the Microprocessor complex, are responsible for cleaving canonical microRNA primary transcripts (pri-miRNAs) to produce microRNA precursors (pre-miRNAs) in animals. This specific cleavage serves as the entrance to the microRNA (miRNA) maturation pathway. In this review, I summarize the current knowledge regarding how Drosha and DGCR8 recognize the pri-miRNA substrates, how pri-miRNA processing may be regulated through a variety of mechanisms, how a heme cofactor binds to DGCR8 and contributes to pri-miRNA processing, and how heterozygous deletion of *DGCR8* causes abnormal miRNA expression and results in neurological defects observed in DiGeorge syndrome. Also discussed are potential functions of DGCR8 and Drosha in regulating the abundance of other RNAs, such as messenger RNAs and long noncoding RNAs.

The Enzymes, Volume 32
ISSN 1874-6047
http://dx.doi.org/10.1016/B978-0-12-404741-9.00005-2

101

1. INTRODUCTION

Canonical microRNAs (miRNAs) are cleaved from primary transcripts (pri-miRNAs) to produce intermediates called precursor miRNAs (pre-miRNAs), which are then further diced into the mature form [1]. The discovery and characterization of miRNAs at these processing stages and the relevant processing factors follow the reverse sequence of the natural pathway. Ambros and Ruvkun described the first miRNA lin-4 and its interaction with a target messenger RNA lin-14 in 1993 [2,3]. Ambros observed the 21-nt mature lin-4 as lin-4S and the 60-nt precursor lin-4 as lin-4L [2]. In 2002, Narry Kim reported the first in-depth characterization of pri-miRNAs [4]. Her group found that miRNAs are transcribed as long and often polycistronic pri-miRNAs, which are processed into pre-miRNAs in the nucleus. For example, pri-miR-30a is an ∼600-nt long RNA as an intergenic independent transcript. miRNAs may arise as independent transcripts, or as parts of messenger RNAs. Some miRNAs reside in the introns and others in the exons.

This review focuses on the first step of the canonical miRNA processing pathway in animals, where pri-miRNAs are recognized and cleaved by the ribonuclease Drosha and its essential RNA-binding partner protein DiGeorge syndrome critical region gene 8 (DGCR8, called Pasha in insects and worms). To understand the functions of miRNAs in biology and diseases, it is critical to understand how Drosha and DGCR8 work together to recognize the several hundreds of pri-miRNAs, to cleave them at precision locations, and how they are regulated through a variety of mechanisms. While canonical miRNAs make up the vast majority of miRNAs, alternative pathways exist. For example, a small number of miRNAs are derived from pre-miRNAs that are produced from introns by the spliceosome and debranching enzymes, instead of the Microprocessor complex [5]. Some small nucleolar RNAs (snoRNAs) may becleaved by Dicer, but not by Drosha, to generate miRNA-like small RNAs [6]. In plants, pri-miRNAs are processed by Dicer-like 1, the double-stranded RNA-binding protein HYPONASTIC LEAVES1, the C2H2-zinc finger protein SERRATE and the nuclear cap-binding complex; the animal and plant systems have some similarity but also large distinctions (reviewed in Ref. [7]).

2. DROSHA

Drosha is a ribonuclease (RNase) III family enzyme. This endonuclease family is characterized by the presence of RNase III domains and double-stranded RNA-binding domains (dsRBDs) [8]. RNase III enzymes cleave double-stranded RNA (dsRNA) substrates at two staggered sites to generate products containing 5′-phosphate, 3′-hydroxyl, and 2-nt 3′-overhangs. The RNase III family can be divided into three subfamilies based on their domain organization. The first subfamily includes bacterial RNase III and yeast Rnt1, which contain one RNase III domain and one or more dsRBDs. They form homodimers to assemble two active sites for the staggered cleavages. The bacterial RNase III is important for processing ribosomal RNA precursors (pre-rRNAs). The Rnt1 in *Saccharomyces cerevisiae* plays an essential role in the maturation of rRNAs, as well as small nuclear RNAs and snoRNAs. Rnt1 also functions in mRNA quality control by cleaving intronic sequences in unspliced mRNAs [9]. The second subfamily is represented by Drosha, which contains two RNase III domains followed by a single dsRBD in the C-terminal region. The third subfamily members are Dicer enzymes that also consist of two RNase III domains and one dsRBD in the C-terminal region, but uniquely include helicase domains, DUF283, and PAZ domains in the N-terminal moiety. As an exception, the Dicer from *Giardia intestinalis* does not possess a helicase domain, a DUF283, or a dsRBD [10]. Dicer is responsible for processing pre-miRNAs to produce miRNA duplexes and for cleaving long dsRNAs to generate small interfering RNAs (siRNAs). The classification of RNase III enzymes in the three subfamilies based on their domain organization and biological functions is not always clear cut. For example, some budding yeast species, including *Saccharomyces castellii* and *Candida albicans*, use noncanonical Dicer proteins to produce siRNAs and mediate RNA interference; these yeast Dicer proteins contain a single RNase III domain, similar to the bacterial RNase III subfamily [11,12]. Recently, the *C. albicans* Dicer was shown to be important not only for RNA interference but also for rRNA maturation [13]. This Enzymes book series contains two excellent chapters in which Rnt1 (Volume 31, Chapter 10) and Dicer (Volume 32, Chapter 1) are discussed.

Drosha contains a proline-rich region, an RS-rich region, a highly conserved central region (with no recognizable motifs), two RNase III domains, and a dsRBD (Fig. 5.1A). In 2000, Crooke and colleagues characterized the human Drosha, before its function in pri-miRNA processing was identified [14]. They found that the Drosha gene is ubiquitously expressed in human tissues and cell lines, and the levels of Drosha mRNA and protein do not change during the cell cycle. Drosha is a nuclear protein, with a fraction further localized to the nucleolus in the S phase during the cell cycle. A recombinant Drosha protein containing the two RNase III domains and the dsRBD cleaves double-stranded RNAs, but not single-stranded RNAs (ssRNAs). Inhibition of Drosha using antisense oligonucleotides increases the levels of pre-rRNAs, suggesting a function in rRNA maturation

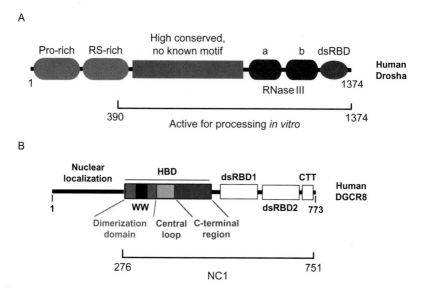

Figure 5.1 Domain structures of Drosha (A) and DGCR8 (B). (See color plate section in the back of the book.)

[14,15]. Indeed, later it was shown that a Drosha-containing complex purified from mouse cells can cleave pre-rRNAs *in vitro* [16]. This complex also comprises DGCR8 and the DEAD-box RNA helicase p68 and p72. p68 and p72 are required for pre-rRNA processing both *in vitro* and *in vivo*. Thus, rRNA maturation is a common function of the bacterial RNase III/yeast Rnt1 and Drosha subfamilies.

In 2003, Kim and colleagues identified Drosha as the pri-miRNA processing enzyme [17]. They first cloned and sequenced pre-miR-30a and revealed its 5′ and 3′ ends, which indicate the two cleavage sites for the missing processing nuclease. The pre-miR-30a folds into a stem loop structure containing a 2-nt overhang at its 3′ end. They further showed via mutagenesis that the double-stranded structure close to the cleavage sites is important for processing, whereas the irregularities such as bulges and internal loops are not essential. The staggered cleavage pattern and the requirement for dsRNA structure in the substrates are characteristic of RNase III family enzymes. There are three RNase III enzymes in humans: in addition to Drosha, L44 is a component of the large subunit of the mitochondrial ribosome; Dicer is the enzyme that cleaves pre-miRNAs to produce miRNA duplexes. Indeed, a Drosha protein (actually a Drosha-containing complex) that was ectopically expressed in human cells and affinity-purified specifically cleaves pri-miRNA *in vitro*. Inhibition of Drosha expression using

RNA interference caused accumulation of pri-miRNAs and concurrent reduction of pre- and mature miRNAs, thus confirmed the role of Drosha in miRNA maturation.

Several key features for the Drosha nuclease have been elucidated in subsequent studies. (a) The proline-rich region and most of the RS-rich region are dispensable for pri-miRNA processing *in vitro*, whereas the C-terminal portion of the RS-rich region, the central region, the two RNase III domains, and the dsRBD are essential [18]. (b) Unlike most other RNase III enzymes, Drosha requires an RNA-binding partner protein DGCR8 for pri-miRNA cleavage activity (see below) [19,20,18,21]. (c) Similar to Dicer, Drosha uses its tandem RNase III domains to form an intramolecular dimer and hence two active sites for the staggered cleavage of pri-miRNAs [18]. (d) The C-terminal dsRBD of Drosha is required for pri-miRNA processing [18]. The structure of this domain has been determined using nuclear magnetic resonance [22]. The structure reveals a class $\alpha\beta\beta\beta\alpha$ dsRBD fold with a unique extended $\alpha1-\beta1$ loop. Sequence features and distribution of charged residues on the surface of this dsRBD are consistent with RNA binding. However, the dsRBD of Drosha does not appear to associate with pri-miRNAs whereas the dsRBD1 of DGCR8 does [23]. The difference in RNA binding may be explained by the dynamic properties of the $\alpha1-\beta1$ and $\beta1-\beta2$ loops [23]. It is possible that the Drosha dsRBD is used for protein–protein interactions, as shown for the dsRBDs in the TRBP and PACT proteins [24,25].

3. DGCR8 AS THE RNA-BINDING PARTNER OF DROSHA

In 2004, DGCR8 was identified as an obligate RNA-binding partner of Drosha for pri-miRNA processing in several parallel studies [19,20,18,21]. Two studies [19,21], led by Hannon and Tuschl's groups, were facilitated by a previously reported yeast two-hybrid screen, which revealed an interaction between *Drosophila* Drosha and an RNA-binding protein [26]. Two others, done in Shiekhattar and Kim's laboratories, utilized biochemical purification of Drosha-containing complexes and mass spectrometry and identified DGCR8, which is the human homolog of the *Drosophila* RNA-binding protein [20,18]. Mammalian DGCR8 and its homolog in *C. elegans* and *Drosophila* are required for pri-miRNA processing [19,20,18,21,27,28]. Importantly, recombinant Drosha and DGCR8 proteins are sufficient for reconstituting pri-miRNA processing *in vitro* [20,18]. Because Drosha and DGCR8 copurify with each other and function in miRNA maturation together, they are collectively called the Microprocessor complex [19,20]. The DGCR8 homologs in *Drosophila* and *C. elegans* are named partner of Drosha (Pasha) [19]. The Microprocessor complex was estimated to have a molecular mass of 400–650 kDa [19,20,18]. Furthermore, a larger Drosha-containing complex isolated from human cells comprises ∼30 other polypeptides, including Ewing's sarcoma gene, the RNA helicase DDX17/p72, hnRNPM4 and nucleolin [20,29]. The large Microprocessor complex is also active in pri-miRNA processing, although to a lower degree [20]. The large Microprocessor may be involved in regulated processing of pri-miRNA processing. DGCR8 is capable of specifically binding pri-miRNAs in the absence of Drosha [30,31] and thus is likely a major contributor to recognition of pri-miRNAs (see Section 6).

The DGCR8 protein has been analyzed using cellular and biochemical methods. DGCR8 is primarily localized in the nucleus, especially in the nucleolus and small foci adjacent to splicing speckles [19,29]. The N-terminal 275 amino acids are responsible for the nuclear localization [32]. DGCR8 is expressed rather ubiquitously; in mouse embryos, DGCR8 is expressed at higher levels in neuroepithelium of primary brain, limb bud, vessels, thymus, and around the palate [33].

Sequence analyses of DGCR8 indicate a WW motif in the central region and two dsRBDs close to the C-terminus (Fig. 5.1B). WW motifs typically contain 30–40 amino acids and are characterized by two conserved Trp residues that are 20–22 amino acids apart. Most characterized WW motifs bind proline-containing peptides [34] and this type of interactions is often found in signaling pathways [35]. However, the WW motif in DGCR8 is important for dimerization and association with a heme cofactor [36,37], as will be discussed in Section 4. The presence of two dsRBDs suggests that a primary function of DGCR8 is to bind RNAs. Indeed, truncated DGCR8 proteins containing only the two dsRBDs and the C-terminal tail (CTT) are active in processing of at least some pri-miRNAs [32,31]. However, these truncations are not as active as the DGCR8 constructs that include the WW motif [38].

A crystal structure of the two DGCR8 dsRBDs has been determined [39]. The structure shows that both dsRBDs adopt the αββα fold similar to other dsRBDs. However, an extra helix (H5, residues 692–700) at the C-terminus of dsRBD2 is packed against both domains, providing a compact overall structure and restraining the relative motion of the two dsRBDs. Mutagenesis results suggest that both dsRBDs contribute to pri-miRNA binding using their second α-helices, hence their RNA-binding surfaces point to distinct directions (see below for further discussion). The dynamic relationship between the two dsRBDs has been further explored using molecular dynamics simulation [40]. The results show that modest motions of the two domains correlate with each other, with the interdomain linker and C-terminal H5 helix serving as the pivot. The concerted motion alters the distance between the two RNA-binding surfaces, possibly allowing DGCR8 to bind and recognize pri-miRNAs with a variety of structural differences.

Drosha and DGCR8 are expected to bind each other as they copurify and function together. Early studies suggest that the WW motif of DGCR8 may associate with the proline-rich region of Drosha [20,21]. This hypothesis is unlikely to hold true because the proline-rich region of Drosha is not required for pri-miRNA processing [18] and the surface that a classic WW domain uses to interact with proline-containing ligands is adopted by DGCR8 as a part of the dimerization interface [36,37]. The Drosha–DGCR8 interaction has been analyzed using a variety of truncation constructs [18,32]. The C-terminal portion of the RS-rich region, the central domain, and the dsRBD of Drosha are required for coimmunoprecipitation (co-IP) with DGCR8 [18]. The co-IP between Drosha and DGCR8 is resistant to RNase A treatment, suggesting that the interaction does not depend on association with pri-miRNAs. The CTT of DGCR8, including residues 739–750, is essential for DGCR8 to bind Drosha [32]. It is not entirely clear whether Drosha and DGCR8 bind each other via direct protein–protein interactions.

4. DGCR8 AS A HEME PROTEIN

My group serendipitously found that the DGCR8 protein binds heme. Heme is the protoporphyrin IX with an iron coordinated at the center (Fig. 5.2A). An active recombinant DGCR8 truncation (called NC1, Fig. 5.1B) expressed in *Escherichia coli* has a yellow/brown color (the darkness depends on the protein concentration) [31]. The absorption spectrum of NC1 displays peaks at 366, 450, and 556 nm, respectively. The cofactor bound to NC1 was confirmed to be heme using a variety of methods, including mass spectrometry and pyridine hemochromagen method. In mass spectrometry analyses, under high-energy denaturing conditions, heme dissociated from NC1 was observed at 616 m/z; whereas under low-energy nondenaturing conditions, dimeric NC1 bound to one heme molecule

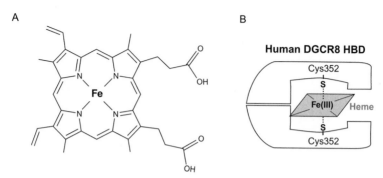

Figure 5.2 DGCR8 is a heme protein. (A) Chemical structure of heme *b* that binds to DGCR8. (B) Schematic of how DGCR8 dimer ligates the Fe(III) in heme using two Cys352 side chains. (For color version of this figure, the reader is referred to the online version of this chapter.)

was the dominant species. Dimerization of NC1 is required for heme binding. The pyridine hemochromagen takes advantage of the characteristic absorption peak at ∼557 nm formed by heme in pyridine and is a classic way to identify heme and measure heme concentrations.

Is the DGCR8–heme interaction an artifact of overexpressing a human protein in *E. coli*, or is heme important for the biological functions of DGCR8? My group investigated this question using a variety of biochemical assays. All the evidence we obtained supports the latter possibility. (a) The heme that is produced by *E. coli* and bound to the recombinant DGCR8 is chemically identical to the heme *b* in humans (Fig. 5.2A). Heme *b* is the major heme isoforms in humans and is the prosthetic cofactor of hemoglobin responsible for transport of oxygen. (b) DGCR8 binds heme using a conserved heme-binding domain located in the central region of this 773-aa protein (Fig. 5.1B). This heme-binding domain can be expressed in *E. coli* as a soluble protein in the absence of other DGCR8 domains, and purified HBD displays electronic absorption spectra similar to that of NC1 [41]. (c) Heme binding is a conserved property of DGCR8 homologs [37]. (d) All heme-binding-deficient DGCR8 mutants, including deletion of the HBD, are defective in pri–miRNA processing in biochemical assays and cultured human cells [31,41] (Sara Weitz, Ming Gong, Shimon Weiss, and Feng Guo, unpublished).

To decipher the biological function of heme in pri–miRNA processing, we characterized the DGCR8–heme interaction using biochemical and crystallographic methods. Our results show that the HBD may be divided into three regions, the N-terminal dimerization domain, the central loop, and the C-terminal region. ESI mass spectrometry analyses under non-denaturing conditions indicate that the N-terminal region of the HBD encode a dimerization domain [41]. In particular, this region contains the WW motif. Most WW motifs fold into monomeric modular domains with a three-stranded β-sheet [34]. Our high-resolution crystal structures of the human and frog DGCR8 dimerization domains show that the WW motif folds into a three-stranded β-sheet, similar to the structure of other WW domains [41,37]. However, the DGCR8 WW motif forms a continuous β-sheet with a fourth strand located at the immediate C-terminal neighboring region of the partner subunit. This "domain swapping" generates an extensive dimerization interface, which is mainly mediated by hydrophobic interactions. Further, all the amino acid residues that are known to be important for heme binding, including Trp329 (the second Trp in the WW motif), Pro351 and Cys352 (the axial ligand of the heme) from both subunits, cluster on a common surface of the dimerization domain. Thus, DGCR8 makes a novel use of the WW motif as a structural platform for dimerization and heme binding.

The central region of the DGCR8 HBD forms a loop (aa 377–410) that is rich in acidic residues and contains a caspase cleavage site [36], as well as a number of predicted phosphorylation sites [42]. This central loop is dispensable for heme binding [36]. The C-terminal region of the HBD is required for heme binding [41]. However, a detailed characterization of this region has not been reported.

Heme proteins often use one or two residues to coordinate (ligate) the heme iron [43]. The coordination contributes to their affinity for heme and is often important for their biological functions. For example, cytochrome P450 enzymes utilize one cysteine side chain to ligate heme iron from one side of the heme plane [44]. The heme iron is coordinated by four nitrogens from the protoporphyrin ring (Fig. 5.2A), thus the cysteine in P450 is called the fifth (proximal) ligand; whereas the position of the sixth (distal) ligand is used for substrate binding and catalytic purposes. DGCR8 is the first example of a heme protein ligating to heme iron using two cysteine side chains [45]. The heme iron in recombinant DGCR8 proteins expressed in bacteria and insect cells is Fe(III) (ferric) [45]. The double-cysteine ligation configuration attributes a strong preference of DGCR8 for Fe(III) over Fe(II) [38]. Importantly, we recently show that Fe(III) heme binds to dimeric apoDGCR8 protein and activates its pri-miRNA processing activity, whereas Fe(II) heme does not [38]. These results clearly demonstrate the importance of DGCR8–heme interaction in miRNA maturation.

Dimerization and heme binding are conserved features of DGCR8 family proteins [37]. Among the three regions of the HBD, the dimerization domain is the most highly conserved, whereas the central loop is the least conserved, and the C-terminal region is modestly conserved. The crystal structure of the frog DGCR8 dimerization domain is almost identical to that of the human protein. Moreover, the residues important for heme binding are conserved. The axial ligand Cys352 and its immediate neighboring residue Pro351 are conserved in all DGCR8 homologs. Trp329, the second Trp in the WW motif, is conserved except in insects and worms, where it is substituted by Ala and His, respectively. In general, the DGCR8 sequences are more conserved in vertebrates than in invertebrates. Heme binding of some DGCR8 homologs has been verified experimentally. The HBD of bat star (a type of starfish) DGCR8 binds heme, even though the amino acid sequence is 41% identical to that of human DGCR8 [37]. This observation indicates that DGCR8 from at least some invertebrates binds heme. The potential for insect and worm Pasha to bind heme is being investigated.

5. DGCR8 IN DIGEORGE SYNDROME

The human *DGCR8* gene was previously found to be located at chromosome 22q11.2, a region that is heterozygously deleted in DiGeorge syndrome [33]. Clinical features of DiGeorge syndrome include cardiovascular, thymic and parathyroid, craniofacial anomalies, as well as learning difficulties, behavioral and psychiatric problems. Deletion of chromosome 22q11.2 is the most frequent chromosomal deletion found in humans with an incidence of 1 in 2000–4000 live births [46]. Most DiGeorge syndrome patients have either a 3-Mb deletion, which includes approximately 60 known genes, or a 1.5-Mb deletion that is nested within the 3-Mb region and contains approximately 35 genes. Both microdeletions comprise *DGCR8*.

Mouse genetics experiments support a contribution of monoallelic deletion of *DGCR8* to DiGeorge syndrome in humans. While carefully examining a mouse strain carrying a hemizygous deficiency $(Df(16)A^{+/-})$ that is syntenic to the 22q11.2 microdeletion in humans, Karayiorgou, Gogos and colleagues observed consistent upregulation of several miRNA-containing transcripts (pri-miRNAs) in both prefrontal cortex and hippocampus, accompanied by downregulation of a subset of mature miRNAs [47]. In the $Df(16)A^{+/-}$ mice, the abundance of *Dgcr8* mRNA is decreased to about 60% that of the wild type, consistent with the inclusion of *Dgcr8* in the deleted region. These observations suggest that a miRNA processing defect may be involved in DiGeorge syndrome. Indeed, *Dgcr8*$^{+/-}$ mice show similar increase in pri-miRNA levels and decrease in mature miRNA expression. The *Dgcr8* heterozygous mice display behavioral and neurological deficits that associate with the 22q11.2 microdeletion, suggesting that the miRNA processing defect caused by decreased expression of *Dgcr8* likely makes important contribution to certain behavior and cognitive DiGeorge symptoms. Subsequent studies further reveal altered short-term plasticity and development of excitatory synaptic transmission in the prefrontal cortex in *Dgcr8*$^{+/-}$ mouse models [48,49].

6. HOW DO DGCR8 AND DROSHA STRUCTURALLY RECOGNIZE PRI-MIRNAS?

There is no sequence common to all pri-miRNAs. Thus, the Microprocessor complex has to recognize structure features of pri-miRNAs. pri-miRNAs contain a hairpin with three helical turns of paired region (Fig. 5.3). Mature miRNAs may be located on either side of the paired region, whereas the opposite strand encodes the passenger strand. Phylogenetic analyses of miRNA genes indicated that the paired strands are highly conserved, some but not all terminal loops are conserved, whereas the regions flanking the hairpin are drastically more variable [50]. This "phylogenetic shadowing" can be used to predict miRNA genes from genomic

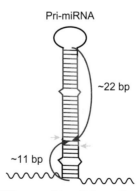

Pri-miRNA

~22 bp

~11 bp

Figure 5.3 Models of pri-miRNA recognition. Schematic of a typical pri-miRNA secondary structure. The red/blue strands represent the mature miRNA and its passenger strand. The yellow arrowheads indicate the Drosha cleavage sites. Two models for pri-miRNA recognition and cleavage site determination are indicated by green arrows. See main text for details. (See color plate section in the back of the book.)

sequences. However, these characteristics of pri-miRNA sequences are not sufficient to computationally define miRNA genes because miRNA hairpins usually contain a number of variations, such as bulges, internal loops, at irregular locations. When these irregularities are allowed in miRNA prediction programs, hundreds of thousands of other sequences become indistinguishable from true miRNAs [51]. It is noteworthy that the MC-fold algorithm for RNA secondary structure, which is based on the knowledge of available three-dimensional RNA structures, predicts that pri-miRNA hairpins are significantly more regular than they appear in the predictions by thermodynamic parameter-based programs [52].

Several important features of pri-miRNAs are revealed by studying their processing in cells or by the Microprocessor complex *in vitro*. The exact sequence of a pri-miRNA is not important for processing [53]. Instead, the pairing interaction in the bottom one-third of the pri-miR-30a hairpin is essential, whereas removal of irregularities is well tolerated [17,54,53]. These observations allow the design of artificial pri-miRNAs that mimic the natural pri-miR-30a to enter the miRNA maturation pathway and efficiently knock down target gene expression [53]. This strategy has been adopted by the second generation of short hairpin RNAs called shRNA[mir] that is widely used in biomedical research [55].

The regions immediately flanking the miRNA hairpin need to be unstructured [56,30]. In *in vitro* pri-miRNA processing assays using affinity-purified Drosha-containing complexes, single-stranded nucleotides on at least one side of the miRNA hairpin are required for efficient processing [56]. The length of the single-stranded region can be as short as 6 nt, but appears to have a dependence on specific pri-miRNAs.

There is a somewhat controversy regarding how Microprocessor specifically recognize pri-miRNAs. An early study using primarily expression of pri-miRNA mutants in 293T cells and some *in vitro* pri-miRNA processing assays suggests that terminal loops in miRNA hairpins need to be at least 10 nt in length, and the Drosha-containing complex determines the cleavage sites by measuring about 22 bp from the terminal loop (Fig. 5.3) [57]. However, later, another study disputed this model by showing that the terminal loop is dispensable for Drosha cleavage, and the Drosha/DGCR8 complex measures about one helical turn from the junction between the miRNA hairpin and the flanking single-stranded regions (Fig. 5.3) [30]. The latter study primarily used purified recombinant Drosha and DGCR8, and further reported that DGCR8 can specifically bind pri-miRNAs as detected via UV-cross-linking. A more recent study reinforced the early conclusion that the terminal loop is important for cleavage by Drosha (Microprocessor) [58]. This study further indicates that the terminal loop also influences the Dicer cleavage step, providing additional information for reconciling results from cellular and biochemical studies.

DGCR8 appears to play a major role in recognition of pri-miRNAs. It is capable of binding pri-miRNAs and discriminating against nonspecific nucleic acids in competition assays even in the absence of Drosha [30,59]. UV cross-linking with pri-miRNAs has been observed for DGCR8, but not for Drosha [30]. A human Drosha fragment containing the RS-rich region (residues 216–333) demonstrates a preference for binding RNA over DNA, but this region is dispensable for pri-miRNA processing *in vitro* [56]. Nevertheless, Drosha must bind pri-miRNAs prior to cleavage. Thus, it is probably reasonable to expect Drosha to contribute to pri-miRNA recognition to certain degree. The lack of understanding of the Drosha–pri-miRNA interaction is largely due to technical difficulties in working with recombinant Drosha proteins.

At least two models have been proposed to explain how DGCR8 binds pri-miRNAs. One model is based on the observation that the RNA-binding surfaces of both DGCR8 dsRBDs contribute to pri-miRNA binding and they point to two different directions [39]. In this model, the three-turn helix of a miRNA hairpin is divided into three roughly equal segments, with the outermost segments contacted by the two dsRBDs and the central segment serving as the linker. Consistent with this model, fluorescence resonance energy transfer analyses suggest that the ends of three-turn helix in a miRNA hairpin become closer to each other upon binding to DGCR8.

My group proposed another model based on the observations obtained using quantitative filter-binding assays and size exclusion chromatography that active recombinant DGCR8 constructs bind pri-miRNAs with high cooperativity (Hill coefficient ≈ 3) and assemble into higher-order complexes (trimer of dimers = hexamer) upon binding pri-miRNAs [31]. By varying DGCR8 concentration in pri-miRNA processing assays, we show that the association of pri-miRNA substrate with DGCR8 strongly correlates with the rate of Drosha-mediated cleavage [59]. Inactive DGCR8 constructs lacking the C-terminal residues (residues 701–773) binds pri-miRNAs with affinity similar to that of the wild-type protein, but with reduced cooperativity (Hill coefficient ≈ 2). These results suggest that these inactive DGCR8 can dimerize upon binding to pri-miRNAs, but fail to add the third dimer subunit to the complex to form a productive complex, competent for triggering cleavage by Drosha. This CTT (amino acid

701–751) contains a predicted amphipathic helix. Mutation of the highly conserved hydrophilic residues on this helix reduces the cooperativity (Hill coefficient ≈ 2) without reducing the affinity for pri-miRNAs, but renders DGCR8 inactive in pri-miRNA processing. Thus, formation of a proper higher-order DGCR8–pri-miRNA complex appears to be critical for processing.

The highly cooperative binding allows DGCR8 to distinguish pri-miRNAs from a nonspecific RNA (the P4–P6 domain of the *Tetrahymena* self-splicing intron [60]), which contains a number of double-stranded and single-stranded regions [59]. The $K_{1/2}$ values of DGCR8 for pri-miR-30a and for P4–P6 are 24 and 50 nM, respectively. However, DGCR8 has no problem distinguished these two RNAs in competition assays. Thus, cooperative binding of DGCR8 to pri-miRNAs is a key mechanism coupling recognition of pri-miRNA to cleavage by the Drosha nuclease. Finally, our negative-stained electron tomography study of the DGCR8–pri-miR-30a complex reveals a complex with dimensions consistent with a hexameric DGCR8 complex associated with one pri-miRNA molecule. To gain a complete understanding of the molecular mechanism through which Drosha and DGCR8 recognize pri-miRNAs, further investigation, especially high-resolutions structures of DGCR8–pri-miRNA binary complexes and Drosha–DGCR8–pri-miRNA tertiary complexes, is needed.

7. HOW IS PRI-MIRNA PROCESSING REGULATED?

Microprocessor autoregulates its own expression via posttranscriptional mechanisms. The *DGCR8* mRNA contains two highly conserved pri-miRNA-like hairpins, one located in the 5′-untranslated region (UTR) and the other in the coding sequence of the N-terminal amino acid residues [61,62]. Two miRNAs, miR-1306-5p and miR-1306-3p, have been detected in several deep sequencing studies. These hairpins are cleaved by the Drosha–DGCR8 complex, and the cleavage reduces the expression of the DGCR8 protein. Further, the DGCR8 protein stabilizes Drosha via protein–protein interaction. The combination of these posttranscriptional mechanisms allows autoregulation of the Microprocessor complex and contributes to the homeostasis of miRNA processing. A similar autoregulation is mediated by a hairpin in the 5′-UTR of *pasha* in Drosophila [63].

Most recently, the functional importance of the Microprocessor autoregulation is demonstrated using mathematical modeling and gene expression analyses [64]. The mathematical modeling addresses the needs for Microprocessor to assure efficient pri-miRNA processing while avoiding off-target cleavage. Good performance, which means high efficiency and high specificity, is achieved only within a narrow range of free Microprocessor concentrations. The autoregulation of Microprocessor described above allows its performance to be maintained with fluctuation of diverse biochemical parameters. Consistent with this theory, the Microprocessor level was found to correlate with the abundance of miRNAs (representing pri-miRNA levels) in different cell types and tissues.

Posttranslational modifications have been demonstrated for DGCR8 and Drosha. Overexpression of histone deacetylase 1 (HDAC1) causes large-scale upregulation of mature miRNAs, but not their corresponding pri-miRNAs [65]. Conversely, decreased expression of HDAC1 reduces the abundance of majority of miRNAs. These observations are consistent with a positive regulation of pri-miRNA processing by HDAC1. Class I HDACs, including HDAC1, HDAC2, and HDAC3, coimmunoprecipitate with Drosha, DGCR8, and the RNA helicases DDX5 and DDX17; whereas class II and class III HDACs are not found to be associated with Drosha. These observations indicate that class I HDACs are components of the large Microprocessor complex. Histone acetyltransferases (HATs), the enzymes that counteract HDACs, decrease the levels of mature miRNAs. HDAC1 increases the pri-miRNA processing activity *in vitro*. Acetylated DGCR8 was detected using an antibody against acetylated lysine. Although the acetylation sites on DGCR8 have not been determined, HAT-treatment of recombinant DGCR8 decreases the affinity for pri-miRNAs and this effect may be reversed via HDAC treatment. Low abundance of HDAC was proposed to be responsible for low levels of miRNA expression in multipotent stem cells and progenitor cells in early stages of development.

Another posttranslational regulation of DGCR8 is mediated via proteolytic cleavage by caspases [36]. DGCR8 expressed in HeLa cells was found to be cleaved in the central loop of the heme-binding domain. Examination of the DGCR8 sequence revealed a conserved caspase cleavage site in the central loop. Indeed, a caspase inhibitor reduces the cleavage of DGCR8 in HeLa cells, and recombinant caspase-3 cleaves DGCR8 *in vitro*. Interestingly, cleavage of DGCR8 by caspases results in separation of the two proteolytic products, dissociation of the heme cofactor, and loss of pri-miRNA processing activity. Decreased miRNA expression in apoptotic cells has been observed [66]. This study, together with an earlier report that Dicer is cleaved and inhibited by caspases [66,67], suggests that both the nuclear and cytoplasmic steps of the miRNA processing pathway are inhibited during apoptosis.

Drosha has been shown to be phosphorylated at Ser300 and Ser302 residues [68]. These residues are located in the region (aa 270–390) responsible for nuclear localization. Phosphorylation of these residues is required for nuclear localization of Drosha and for pri-miRNA processing. The glycogen synthase kinase 3 beta (GSK3β), a cytoplasmic serine/threonine kinase, is responsible for phosphorylation of Drosha at these two positions [69]. Inhibition of GSK3β using a small molecule inhibitor causes significant amount of Drosha to be mislocalized to the cytoplasm. GSK3β is involved in inflammation, cell proliferation and migration, apoptosis; it serves as an important component of the Wnt signaling pathway and is targeted for therapeutics of several diseases including type II diabetes, Alzheimer's disease, cancer, and bipolar disorder [70]. Thus, GSK3β-mediated phosphorylation of Drosha and alteration of miRNA maturation may be involved in these biological processes and should be considered when evaluating the effects of GSK3β inhibitors under normal and disease conditions.

Increasing number of proteins, such as the DEAD-box RNA helicases p68 and p72 [16], p53 [71], SMAD [72], KSRP [73], lin28 [74,75], have been demonstrated to regulate the processing of subsets of miRNAs. In general, these proteins bind pri-miRNAs and either recruit Drosha/DGCR8 to promote the processing or inhibit the processing. Some of them are auxiliary components of the larger Microprocessor complex [20,16]. The subject of how the auxiliary proteins regulate pri-miRNA processing is nicely reviewed in Chapters 6 and 8, and will not be elaborated here.

8. DROSHA AND DGCR8 BIND RNAS OTHER THAN PRI-miRNAs

The observations that the Microprocessor complex cleaves the *DGCR8* mRNA and about 10% of miRNAs are located in coding regions of mRNAs raise the possibility that Microprocessor controls the stability of other mRNAs. This theory has gained substantial support, but not without skepticism. In HeLa cells, about 100 mRNAs are upregulated when Drosha or DGCR8 is downregulated using RNA interference, but not when Dicer or Ago2 is knocked down [61]. Similarly, in *Drosophila* S2 cells, more than 100 RNAs that are not miRNAs are regulated by Drosha but not by Dicer-1, as demonstrated via RNA interference and microarray analyses [63].

However, an analysis of deep sequencing data for small RNAs in the size ranges covering mature miRNAs and pre-miRNAs did not yield support for Microprocessor-mediated cleavage and regulation of other mRNAs [76].

Recently, an analysis of DGCR8-associated RNAs in HEK 293T cells using high-throughput sequencing and cross-linking immunoprecipitation revealed several hundred mRNAs, snoRNAs, and long noncoding RNAs (lncRNAs) [77]. The experiments were performed using either endogenous or overexpressed DGCR8 and similar results were obtained. Among ~8000 targets common to both endogenous and overexpressed DGCR8, 43% are protein coding genes, including 26% introns, 12% amino acid coding sequences, and 5% UTRs. The binding and regulation by DGCR8 have been confirmed for at least several mRNAs. Further, 241 DGCR8-binding sites are cassette exons, suggesting an interesting possibility that DGCR8 may affect the relative abundance of alternatively spliced mRNAs. Regulation of several such mRNAs has been confirmed by comparing wild-type, $Dgcr8^{-/-}$ and $Dicer^{-/-}$ cells. DGCR8 negatively affects the abundance of snoRNAs such as U16 and U92 in a Drosha-independent fashion. This observation indicates that DGCR8 and Drosha do not always function together. There might be other currently unknown endonucleases that work with DGCR8 to cleave snoRNAs. Finally, one of the DGCR8-bound lncRNAs is *MALAT1* (metastasis-associated lung adenocarcinoma transcript 1). Microprocessor cleaves *MALAT1* and negatively regulates its abundance. Overall, this study provides strong support for functions of DGCR8 and Drosha beyond pri-miRNA processing and opens the door for further investigation of these two proteins in normal biological processes and in diseases.

9. FUTURE PERSPECTIVES

Despite the substantial knowledge that has been accumulated, many key questions regarding how Drosha and DGCR8 function remain unanswered. (a) How are pri-miRNAs recognized by Drosha and DGCR8? In addition to resolving the controversial models regarding pri-miRNA recognition, we especially need to identify the protein moieties that recognize the ssRNA–dsRNA junctions in pri-miRNAs. (b) How is the nuclease activity of Drosha triggered by pri-miRNAs and DGCR8? (c) How does the Microprocessor complex release the products and thus turnover in successive pri-miRNA processing? (d) It is thought that RNAs are never truly "free" between different processing steps as often depicted in simplified diagrams. Do Drosha and DGCR8 actively transfer pre-miRNAs to the next step, which is presumably the Exportin-5-mediated nuclear export? (e) How do DGCR8 and Drosha regulate the abundance of other cognate RNAs? (f) How do the auxiliary factors interact with pri-miRNAs, Drosha, or DGCR8 and regulate processing? Answers to these questions will not only provide an understanding of these fundamental biological processes, but also help to improve the miRNA gene prediction programs, to potentially enhance the design of shRNAs, to understand the role of altered pri-miRNA processing in diseases, and to suggest potential ways to modulate pri-miRNA processing using small molecular inhibitors or activators.

ACKNOWLEDGMENTS

I apologize for any relevant references that are not cited due to limited space and emphases of the discussion. This work is supported by a National Institutes of Health (NIH) R01 grant (GM080563).

REFERENCES

[1] Kim VN, Han J, Siomi MC. Biogenesis of small RNAs in animals. Nat Rev Mol Cell Biol 2009;10:126–39.
[2] Lee RC, Feinbaum RL, Ambros V. The *C. elegans* heterochronic gene lin-4 encodes small RNAs with antisense complementarity to lin-14. Cell 1993;75:843–54.
[3] Wightman B, Ha I, Ruvkun G. Posttranscriptional regulation of the heterochronic gene lin-14 by lin-4 mediates temporal pattern formation in *C. elegans*. Cell 1993;75:855–62.
[4] Lee Y, Jeon K, Lee JT, Kim S, Kim VN. MicroRNA maturation: stepwise processing and subcellular localization. EMBO J 2002;21:4663–70.
[5] Westholm JO, Lai EC. Mirtrons: microRNA biogenesis via splicing. Biochimie 2011;93:1897–904.
[6] Scott MS, Ono M. From snoRNA to miRNA: dual function regulatory non-coding RNAs. Biochimie 2011;93:1987–92.
[7] Voinnet O. Origin, biogenesis, and activity of plant microRNAs. Cell 2009;136:669–87.
[8] MacRae IJ, Doudna JA. Ribonuclease revisited: structural insights into ribonuclease III family enzymes. Curr Opin Struct Biol 2007;17:138–45.
[9] Danin-Kreiselman M, Lee CY, Chanfreau G. RNAse III–mediated degradation of unspliced pre-mRNAs and lariat introns. Mol Cell 2003;11:1279–89.
[10] MacRae IJ, et al. Structural basis for double-stranded RNA processing by Dicer. Science 2006;311:195–8.
[11] Drinnenberg IA, et al. RNAi in budding yeast. Science 2009;326:544–50.
[12] Weinberg DE, Nakanishi K, Patel DJ, Bartel DP. The inside-out mechanism of Dicers from budding yeasts. Cell 2011;146:262–76.
[13] Bernstein DA, Vyas VK, Weinberg DE, Drinnenberg IA, Bartel DP, Fink GR. Candida albicans Dicer (CaDcr1) is required for efficient ribosomal and spliceosomal RNA maturation. Proc Natl Acad Sci USA 2012;109:523–8.
[14] Wu H, Xu H, Miraglia LJ, Crooke ST. Human RNase III is a 160-kDa protein involved in preribosomal RNA processing. J Biol Chem 2000;275:36957–65.

[15] Liang XH, Crooke ST. Depletion of key protein components of the RISC pathway impairs pre-ribosomal RNA processing. Nucleic Acids Res 2011;39:4875–89.

[16] Fukuda T, et al. DEAD-box RNA helicase subunits of the Drosha complex are required for processing of rRNA and a subset of microRNAs. Nat Cell Biol 2007;9:604–11.

[17] Lee Y, et al. The nuclear RNase III Drosha initiates microRNA processing. Nature 2003;425:415–9.

[18] Han J, Lee Y, Yeom KH, Kim YK, Jin H, Kim VN. The Drosha-DGCR8 complex in primary microRNA processing. Genes Dev 2004;18:3016–27.

[19] Denli AM, Tops BB, Plasterk RH, Ketting RF, Hannon GJ. Processing of primary microRNAs by the Microprocessor complex. Nature 2004;432:231–5.

[20] Gregory RI, et al. The Microprocessor complex mediates the genesis of microRNAs. Nature 2004;432:235–40.

[21] Landthaler M, Yalcin A, Tuschl T. The human DiGeorge syndrome critical region gene 8 and its *D. melanogaster* homolog are required for miRNA biogenesis. Curr Biol 2004;14:2162–7.

[22] Mueller GA, Miller MT, Derose EF, Ghosh M, London RE, Hall TM. Solution structure of the Drosha double-stranded RNA-binding domain. Silence 2010;1:2.

[23] Wostenberg C, Quarles KA, Showalter SA. Dynamic origins of differential RNA binding function in two dsRBDs from the miRNA "microprocessor" complex. Biochemistry 2010;49:10728–36.

[24] Gupta V, Huang X, Patel RC. The carboxy-terminal, M3 motifs of PACT and TRBP have opposite effects on PKR activity. Virology 2003;315:283–91.

[25] Haase AD, et al. TRBP, a regulator of cellular PKR and HIV-1 virus expression, interacts with Dicer and functions in RNA silencing. EMBO Rep 2005;6:961–7.

[26] Giot L, et al. A protein interaction map of Drosophila melanogaster. Science 2003;302:1727–36.

[27] Wang Y, Medvid R, Melton C, Jaenisch R, Blelloch R. DGCR8 is essential for microRNA biogenesis and silencing of embryonic stem cell self-renewal. Nat Genet 2007;39:380–5.

[28] Yi R, et al. DGCR8-dependent microRNA biogenesis is essential for skin development. Proc Natl Acad Sci USA 2009;106:498–502.

[29] Shiohama A, Sasaki T, Noda S, Minoshima S, Shimizu N. Nucleolar localization of DGCR8 and identification of eleven DGCR8-associated proteins. Exp Cell Res 2007;313:4196–207.

[30] Han J, et al. Molecular basis for the recognition of primary microRNAs by the Drosha-DGCR8 complex. Cell 2006;125:887–901.

[31] Faller M, Matsunaga M, Yin S, Loo JA, Guo F. Heme is involved in microRNA processing. Nat Struct Mol Biol 2007;14:23–9.

[32] Yeom KH, Lee Y, Han J, Suh MR, Kim VN. Characterization of DGCR8/Pasha, the essential cofactor for Drosha in primary miRNA processing. Nucleic Acids Res 2006;34:4622–9.

[33] Shiohama A, Sasaki T, Noda S, Minoshima S, Shimizu N. Molecular cloning and expression analysis of a novel gene DGCR8 located in the DiGeorge syndrome chromosomal region. Biochem Biophys Res Commun 2003;304:184–90.

[34] Sudol M. The WW domain. In: Cesareni G, Gimona M, Sudol M, Yaffe M, editors. Modular protein domains. Weinheim, Germany: Wiley-VCH; 2005. p. 59–72.

[35] Sudol M, Harvey KF. Modularity in the Hippo signaling pathway. Trends Biochem Sci 2010;35:627–33.

[36] Gong M, Chen Y, Senturia R, Ulgherait M, Faller M, Guo F. Caspases cleave and inhibit the microRNA processing protein DiGeorge Critical Region 8. Protein Sci 2012;21:797–808.

[37] Senturia R, Laganowsky A, Barr I, Scheidemantle BD, Guo F. Dimerization and heme binding are conserved in amphibian and starfish homologues of the microRNA processing protein DGCR8. PLoS One 2012;7:e39688.

[38] Barr I, Smith AT, Chen Y, Senturia R, Burstyn JN, Guo F. Ferric, not ferrous, heme activates RNA-binding protein DGCR8 for primary microRNA processing. Proc Natl Acad Sci USA 2012;109:1919–24.

[39] Sohn SY, Bae WJ, Kim JJ, Yeom KH, Kim VN, Cho Y. Crystal structure of human DGCR8 core. Nat Struct Mol Biol 2007;14:847–53.

[40] Wostenberg C, Noid WG, Showalter SA. MD simulations of the dsRBP DGCR8 reveal correlated motions that may aid pri-miRNA binding. Biophys J 2010;99:248–56.

[41] Senturia R, et al. Structure of the dimerization domain of DiGeorge Critical Region 8. Protein Sci 2010;19:1354–65.

[42] Rost B, Yachdav G, Liu J. The PredictProtein server. Nucleic Acids Res 2004;32: W321–W326.

[43] Paoli M, Marles-Wright J, Smith A. Structure-function relationships in heme-proteins. DNA Cell Biol 2002;21:271–80.

[44] Ortiz de Montellano PR. Cytochrome P450: structure, mechanism, and biochemistry. 3rd ed. New York, NY: Springer; 2004.

[45] Barr I, et al. DiGeorge Critical Region 8 (DGCR8) is a double-cysteine-ligated heme protein. J Biol Chem 2011;286:16716–25.

[46] Karayiorgou M, Simon TJ, Gogos JA. 22q11.2 microdeletions: linking DNA structural variation to brain dysfunction and schizophrenia. Nat Rev Neurosci 2010;11:402–16.

[47] Stark KL, et al. Altered brain microRNA biogenesis contributes to phenotypic deficits in a 22q11-deletion mouse model. Nat Genet 2008;40:751–60.

[48] Fenelon K, et al. Deficiency of Dgcr8, a gene disrupted by the 22q11.2 microdeletion, results in altered short-term plasticity in the prefrontal cortex. Proc Natl Acad Sci USA 2011;108:4447–52.

[49] Schofield CM, Hsu R, Barker AJ, Gertz CC, Blelloch R, Ullian EM. Monoallelic deletion of the microRNA biogenesis gene Dgcr8 produces deficits in the development of excitatory synaptic transmission in the prefrontal cortex. Neural Dev 2011;6:11.

[50] Berezikov E, Guryev V, van de Belt J, Wienholds E, Plasterk RH, Cuppen E. Phylogenetic shadowing and computational identification of human microRNA genes. Cell 2005;120:21–4.

[51] Lim LP, Glasner ME, Yekta S, Burge CB, Bartel DP. Vertebrate microRNA genes. Science 2003;299:1540.

[52] Parisien M, Major F. The MC-Fold and MC-Sym pipeline infers RNA structure from sequence data. Nature 2008;452:51–5.

[53] Zeng Y, Wagner EJ, Cullen BR. Both natural and designed micro RNAs can inhibit the expression of cognate mRNAs when expressed in human cells. Mol Cell 2002;9:1327–33.

[54] Zeng Y, Cullen BR. Sequence requirements for micro RNA processing and function in human cells. RNA 2003;9:112–23.

[55] Silva JM, et al. Second-generation shRNA libraries covering the mouse and human genomes. Nat Genet 2005;37:1281–8.

[56] Zeng Y, Cullen BR. Efficient processing of primary microRNA hairpins by Drosha requires flanking nonstructured RNA sequences. J Biol Chem 2005;280:27595–603.

[57] Zeng Y, Yi R, Cullen BR. Recognition and cleavage of primary microRNA precursors by the nuclear processing enzyme Drosha. EMBO J 2005;24:138–48.

[58] Zhang X, Zeng Y. The terminal loop region controls microRNA processing by Drosha and Dicer. Nucleic Acids Res 2010;38:7689–97.

[59] Faller M, et al. DGCR8 recognizes primary transcripts of microRNAs through highly cooperative binding and formation of higher-order structures. RNA 2010;16:1570–83.

[60] Juneau K, Podell E, Harrington DJ, Cech TR. Structural basis of the enhanced stability of a mutant ribozyme domain and a detailed view of RNA–solvent interactions. Structure 2001;9:221–31.

[61] Han J, et al. Posttranscriptional crossregulation between Drosha and DGCR8. Cell 2009;136:75–84.

[62] Triboulet R, Chang HM, Lapierre RJ, Gregory RI. Post-transcriptional control of DGCR8 expression by the Microprocessor. RNA 2009;15:1005–11.

[63] Kadener S, et al. Genome-wide identification of targets of the drosha-pasha/DGCR8 complex. RNA 2009;15:537–45.

[64] Barad O, et al. Efficiency and specificity in microRNA biogenesis. Nat Struct Mol Biol 2012;19:650–2.

[65] Wada T, Kikuchi J, Furukawa Y. Histone deacetylase 1 enhances microRNA processing via deacetylation of DGCR8. EMBO Rep 2012;13:142–9.

[66] Ghodgaonkar MM, et al. Abrogation of DNA vector-based RNAi during apoptosis in mammalian cells due to caspase-mediated cleavage and inactivation of Dicer-1. Cell Death Differ 2009;16:858–68.

[67] Matskevich AA, Moelling K. Stimuli-dependent cleavage of Dicer during apoptosis. Biochem J 2008;412:527–34.

[68] Tang X, Zhang Y, Tucker L, Ramratnam B. Phosphorylation of the RNase III enzyme Drosha at Serine300 or Serine302 is required for its nuclear localization. Nucleic Acids Res 2010;38:6610–9.

[69] Tang X, Li M, Tucker L, Ramratnam B. Glycogen synthase kinase 3 beta (GSK3beta) phosphorylates the RNAase III enzyme Drosha at S300 and S302. PLoS One 2011;6: e20391.

[70] Wu D, Pan W. GSK3: a multifaceted kinase in Wnt signaling. Trends Biochem Sci 2010;35:161–8.

[71] Suzuki HI, Yamagata K, Sugimoto K, Iwamoto T, Kato S, Miyazono K. Modulation of microRNA processing by p53. Nature 2009;460:529–33.

[72] Davis BN, Hilyard AC, Lagna G, Hata A. SMAD proteins control DROSHA-mediated microRNA maturation. Nature 2008;454:56–61.

[73] Trabucchi M, et al. The RNA-binding protein KSRP promotes the biogenesis of a subset of microRNAs. Nature 2009;459:1010–4.

[74] Viswanathan SR, Daley GQ, Gregory RI. Selective blockade of microRNA processing by Lin28. Science 2008;320:97–100.

[75] Piskounova E, et al. Lin28A and Lin28B inhibit let-7 microRNA biogenesis by distinct mechanisms. Cell 2011;147:1066–79.

[76] Shenoy A, Blelloch R. Genomic analysis suggests that mRNA destabilization by the microprocessor is specialized for the auto-regulation of Dgcr8. PLoS One 2009;4: e6971.

[77] Macias S, Plass M, Stajuda A, Michlewski G, Eyras E, Caceres JF. DGCR8 HITS-CLIP reveals novel functions for the Microprocessor. Nat Struct Mol Biol 2012;19:760–6.

CHAPTER SIX

Control of Drosha-Mediated MicroRNA Maturation by Smad Proteins

Hara Kang*,†, Akiko Hata†,1

*Division of Life Sciences, College of Life Sciences and Bioengineering, University of Incheon, Incheon, Republic of Korea
†Cardiovascular Research Institute, University of California at San Francisco, San Francisco, California, USA
1Corresponding author: e-mail address: akiko.hata@ucsf.edu

Contents

Abstract

MicroRNAs (miRNAs) are small ~22 nucleotides (nt) noncoding RNAs that regulate gene expression at the posttranscriptional level. The miRNA biogenesis pathway comprises transcription by RNA polymerase II (Pol II), followed by sequential cropping of primary transcripts performed by two ribonuclease III (RNase III) enzymes, Drosha and Dicer. Regulation of each step of miRNA biogenesis is critical for generating functional mature miRNAs. Recently, Smad proteins, the signal transducers of the TGFβ signaling pathway, have been found to modulate miRNA biosynthesis of a group of miRNAs. In this chapter, we describe our recent understanding of the regulation of Drosha-mediated miRNA processing by Smad proteins.

1. miRNA BIOGENESIS

The human genome encodes over 1000 microRNAs (miRNAs), which are predicted to target about 60% of mammalian genes. Approximately 50% of miRNAs are encoded in noncoding transcripts and about 40% are derived from the introns of protein-coding genes [1,2]. The majority of miRNAs are transcribed by polymerase II (Pol II) as long primary-miRNA transcripts (pri-miRNA) containing a single or multiple hairpin structures [3,4]. Similar to mRNA, the pri-miRNA bear a 7-methylguanylate cap at the 5′ end and a poly (A) tail at the 3′ end [5]. The pri-miRNA hairpin includes an imperfect stem of approximately 30 bp with flanking single-strand RNA (ssRNA) segments at its base, termed the ssRNA–double-stranded RNA (dsRNA) junction [6,7]. This structure is recognized and cleaved by the Drosha Microprocessor complex, which consists of the Drosha RNase III (ribonuclease III) enzyme, the RNA-binding protein DiGeorge syndrome critical region gene 8 (DGCR8), and other cofactors [8–11]. DGCR8 binds to the ssRNA–dsRNA junction of the pri-miRNA hairpins and functions as both a molecular anchor and a ruler: it stabilizes the association of Drosha with the pri-miRNA and promotes cleavage of the pri-miRNAs by Drosha at a distance of approximately 11 nt from the junction. Drosha cleavage occurs cotranscriptionally before splicing of the host RNA, and generates a truncated hairpin of approximately 60 nt precursor-miRNA (pre-miRNA) with a 2 nt 3′-overhang, characteristic of RNase III-mediated cleavage. After the Drosha-mediated processing step, the pre-miRNA is exported from nucleus to cytoplasm through Exportin-5 (Xpo5) in cooperation with the guanine triphosphatase Ran. Xpo5 recognizes the 3′-overhang and interacts with the stem-loop structure of the pre-miRNA. In the cytoplasm, the pre-miRNA associates with the cytoplasmic RNase III enzyme Dicer, which cleaves the pre-miRNA into an approximately 22 nt-long double-stranded miRNA (miRNA duplex) containing the guide strand (mature miRNA) and the passenger strand (miRNA⋆). Dicer recognizes the 3′-end of the pre-miRNA and cleaves two helical turns away in cooperation with HIV-1 transactivation response RNA-binding protein and possibly other proteins, generating a miRNA–miRNA⋆ duplex having 2 nt 3′-overhangs at both ends [12,13]. The miRNA duplex is then loaded into argonaute proteins, which select the guide strand and present it to the RNA-induced silencing complex (RISC) for targeting mRNAs, followed by gene silencing, while the passenger strand is released and degraded

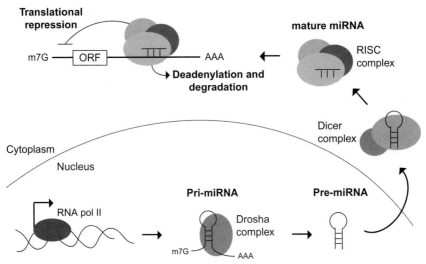

Figure 6.1 Schematic diagram of miRNA biogenesis. Pri-miRNAs are transcribed by RNA Pol II and processed into hairpin-structured pre-miRNAs by Drosha complex. The pre-miRNAs are then exported from the nucleus to the cytoplasm and are cleaved by the Dicer complex into miRNA duplexes. The mature miRNA is incorporated into the RISC complex and mediates posttranscriptional repression of target mRNA by translational repression and/or deadenylation and degradation. (See color plate section in the back of the book.)

(Fig. 6.1). The thermodynamic stability of the ends of the miRNA duplex is thought to contribute to the selection of a single miRNA strand [14–16]. The mature miRNA guides RISC to partially complementary sequences within the target mRNAs to mediate the repression of targets. While most miRNAs are generated through this general pathway, some miRNAs are produced by a noncanonical biogenesis pathway. For example, there are miRNAs termed "mirtrons" located within short introns. The ends of the pre-miRNA hairpin are determined by the splice sites of the intron [17,18]. Following the completion of splicing, the lariat-debranching enzyme resolves the branch point to generate a pre-miRNA-like hairpin that can be exported from the nucleus and processed by Dicer through the canonical miRNA biogenesis pathway. Unlike miRNAs, the sequences of mirtrons are not evolutionarily conserved. Moreover, mammalian mirtrons generally contain single nucleotide overhangs on both strands, while fly and worm mirtrons generate hairpins with 2 nt 3′-overhangs consistent with the structure of canonical miRNAs [19]. In general, mirtrons are expressed at lower level than miRNAs generated by Drosha-dependent processing [19].

2. DROSHA AND DGCR8

Drosha was the first human RNase III enzyme identified and cloned [20]. RNase III enzymes are a family of dsRNA specific ribonucleases that generate staggered cuts on each side of the RNA helix [21]. RNase III enzymes typically contain both RNase III domains and dsRNA-binding domains (dsRBD) and are grouped into three classes based on domain organization [22–24] (Fig. 6.2). Class I enzymes, found in bacteria and yeasts, contain one conserved RNase III domain and an adjacent dsRBD. Class II RNase III enzymes, including Drosha, have tandem RNase III domains and one dsRBD. Class III enzymes, including Dicer, contain a putative helicase domain and the PAZ domain, in addition to tandem RNase III domains and a dsRBD. Sequence analyses and structural studies of RNase III enzymes have revealed that two catalytic domains form a dimer critical for the catalytic function [10,25].

Multiple roles have been reported for RNase III enzymes. In general, they are involved in ribosomal RNA (rRNA) maturation and mRNA degradation. For example, the yeast RNase III enzyme Rnt1p controls nuclear processing of rRNAs, small nuclear RNAs, and snoRNAs, and also plays a role in the degradation of unspliced pre-mRNAs [26]. In humans, it has been suggested that the RNase III enzyme Drosha promotes rRNA processing, as pre-rRNAs accumulates when Drosha is depleted. More

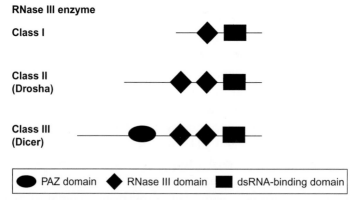

Figure 6.2 Schematic diagram of RNase III enzymes. RNase III enzymes are grouped into three classes based on domain organization. Representative domain structures such as RNase III domain and dsRNA-binding domain are indicated.

recently, it has been found that Drosha plays an essential role in miRNA processing [10]. Immunopurified Drosha cleaves pri-miRNA to generate pre-miRNA *in vitro*. Moreover, knockdown of Drosha results in the accumulation of pri-miRNAs and the consequent reduction of both pre-miRNAs and mature miRNAs *in vivo* [10].

Drosha recognizes pri-miRNA hairpins of a specific structure and length and cleaves to release the pre-miRNA. The sequences flanking the stem loop and/or the terminal loop of pri-miRNA are shown to be important for efficient pri-miRNA processing [7,27]. Drosha is assembled in a large ~ 600 kDa protein complex known as the "Drosha Microprocessor complex" [8–10]. DGCR8, known as "Pasha" in *Drosophila*, is one of its essential subunits. A heterotetramer of two Drosha and two DGCR8 molecules appears to be the critical core of the Microprocessor complex. DGCR8 is thought to bind directly to the central region and the RNase III domains of Drosha [10]. Importantly, DGCR8 protein contains two consensus dsRBD; it binds to the double-stranded stem near the loop of pri-miRNAs. Considering that terminal structure of pri-miRNA influences the orientation of DGCR8 binding to pri-miRNA, pri-miRNAs may allow the Microprocessor to bind more favorably in the productive orientation [7].

3. MODULATORS OF THE DROSHA MICROPROCESSOR COMPLEX

The interaction of Drosha with its cofactors in the Microprocessor complex is crucial for the catalytic activity, specificity, and fidelity of Drosha cleavage. *In vitro* processing assays with purified Drosha and DGCR8 show that, although the many miRNAs can be processed, the "pri-to-pre" cleavage of some miRNAs is relatively slow and inefficient [28]. The DEAD-box RNA helicases p68 (DDX5) and p72 (DDX17) have been identified as components of the Drosha Microprocessor complex [9,29]. Depletion of p68 or p72 results in a reduction in the levels of a large number of miRNAs, suggesting that p68 and p72 play a part in promoting Drosha cleavage of a substantial subset of miRNAs [30]. An interesting recent development sees p68 and p72 as potential molecular bridges linking Drosha to a group of interacting proteins with regulatory functions. Among these, p53, estradiol-bound estrogen receptor-α (ER-α), and Smad proteins, all of which interact with p68 or p72 in response to extracellular stimuli, have been recently implicated in the regulation of Drosha-dependent processing of specific pri-miRNAs [31–34].

In response to DNA damage, the tumor suppressor protein p53 associates with p68 and Drosha on the pri-miRNA and promotes pri-to-pre processing of multiple miRNAs, including miR-143, -145, and -16 [31]. It is likely that this property of p53 might be relevant to cancer biology, since transcriptionally inactive p53 mutants found in various tumors are unable to interact with the Drosha complex and p68, resulting in an attenuation of Drosha-mediated pri-miRNA processing [31].

In contrast to p53, ER-α attenuates pri-to-pre-miRNA processing through association with p68 and Drosha [32,35]. Upon estradiol binding, activated ER-α is recruited to the Drosha complex through p68. It hastens the inactivation of the Microprocessor or its dissociation from the pri-miRNAs, thereby inhibiting the processing of several pri-miRNAs, such as miR-125a, -195, -143, -145, and -16, which in turn results in the upregulation of ER-α target gene transcripts.

In addition to their canonical role in the regulation of transcription in response to ligands of the transforming growth factor β (TGFβ) family, Smads also regulate posttranscriptional processing of miRNAs by Drosha [34] (Fig. 6.3). For example, the level of mature miR-21 is increased upon TGFβ or bone morphogenetic protein 4 (BMP4) stimulation. Induction of miR-21 is resistant to Pol II transcription inhibitors, such as actinomycin D and α-amanitin, suggesting that miR-21 is posttranscriptionally induced. Indeed, upon TGFβ or BMP4 stimulation, Smads associate with the Drosha Microprocessor complex via p68 and facilitate the processing of pri-miR-21 into pre-miR-21. As a consequence, miR-21 reduces the expression of its target genes, contributing to the biological effects of TGFβ or BMP signaling. We will discuss this process in further detail below.

4. SMAD PROTEINS

Smads are signal transducers of the TGFβ family of growth factors, a group of related and evolutionarily conserved cytokines that regulate various biological processes, such as cell differentiation, proliferation, and migration in embryos and adults [36]. TGFβ and BMP are the two major subfamilies of this large lineage [37]. In response to binding of diverse TGFβ-family ligands, a membrane-bound complex of two molecules of type I and two molecules of type II TGFβ receptors, all of them serine/threonine kinases, become activated [38]. After the formation of this heterotetrameric receptor

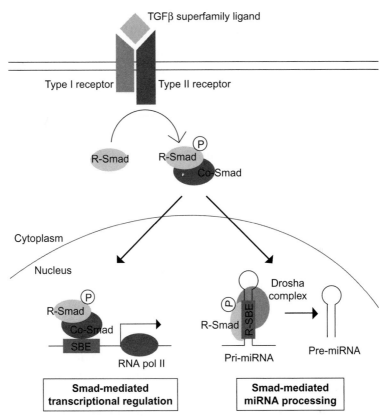

Figure 6.3 Regulation of miRNA by TGFβ superfamily signaling. Stimulation with TGFβ family of ligands results in phosphorylation and translocation of Smads into the nucleus. Smads modulate gene transcription by binding to SBE. Alternatively, Smads directly associate with R-SBE of pri-miRNA and facilitate a cleavage by Drosha. (See color plate section in the back of the book.)

complex, the constitutively active type II receptor kinase phosphorylates the glycine/serine-rich domain of the type I receptor. This phosphorylation event activates the catalytic activity of the type I receptor, and thus propagates the downstream signal by phosphorylation of the Smad signal transducers, which in turn migrate to the nucleus. Three classes of Smads have been identified: receptor-specific Smads (R-Smads), common Smads (co-Smads), and inhibitory Smads [39]. In the canonical signaling pathway, the active receptor complex phosphorylates R-Smads [40]. In particular, TGFβ phosphorylates Smad2 and Smad3 while BMPs activate Smad1, Smad 5, and Smad 8. The phosphorylation of these R-Smads promotes their association with the co-Smad, Smad4, in a complex that translocates to the nucleus and regulates gene transcription, either positively or negatively. Transcriptional regulation occurs in conjunction with other transcription factors through direct binding of a DNA sequence known as the "Smad-binding element (SBE)" in the promoter region of TGFβ target genes [41] (Fig. 6.3).

5. miRNA REGULATION BY SMADS

5.1. Transcriptional regulation of miRNAs by Smads

Nucleosome positioning and chromatin immunoprecipitation analyses of the promoter regions of miRNA genes suggest that miRNA promoter elements are indistinguishable from those of protein-coding genes [4,42]. Therefore, it is not surprising that Smads, as DNA binding factors, bind to promoter regions and modulate the transcription of a cohort of miRNA genes (Fig. 6.3). For example, miR-216a and -217 in glomerular mesengial cells and miR-192 in mouse kidney epithelial cells are transcriptionally induced by Smads [43,44]. Similarly, let-7a and let-7d in human alveolar basal epithelial cells, and miR-155 in murine mammary gland epithelial cells, are induced by TGFβ-mediated Smad binding to SBE in their promoters [45]. Conversely, TGFβ represses transcription of miR-24 in myoblasts and miR-29 in fibroblasts and tubular epithelial cells by Smad binding to their promoters [46,47].

In addition to direct association of the Smad proteins with the miRNA promoters, Smads can indirectly modulate miRNA expression through activation of transcription factors that associate with the promoters of miRNAs. For example, expression levels of miR-143 and -145 are increased by TGFβ/BMP-mediated activation of transcription factors Myocardin and myocardin related transcription factor [48,49].

miRNA promoters are also epigenetically modulated like protein-coding gene promoters. Many miRNA loci, including miR-9-1, -34b/c, -137, -193a, -203, and -304, were found to be hypermethylated in multiple human cancer cells [50,51], while the let-7a locus is hypomethylated in lung adenocarcinoma [52]. Consistently, histone modifications affect miRNAs expression. For instance, inhibition of histone deacetylase (HDAC) activities increases the expression of several miRNAs in cancer cells [53,54]. As Smads recruit HDAC or histone acetyltransferase to the promoters of multiple genes, changes in expression of some miRNAs caused by Smads are possibly due to epigenetic regulation through Smad-mediated histone modification [55].

5.2. Drosha-dependent processing regulation by Smads

A novel role for Smads in the regulation of miRNA processing has been reported recently [34]. TGFβ and BMP4 play a critical role in the regulation of vascular smooth muscle cell (VSMC) phenotype. They elevate expression of genes that encode the contractile apparatus while also decreasing migration and proliferation of VSMCs, thus producing a switch to a differentiated state termed "contractile phenotype." Interestingly, mature miR-21 is induced by TGFβ and BMP4 in VSMCs without a corresponding increase in pri-miR-21, unlike transcriptionally regulated miRNAs. The increased level of miR-21 represses the expression of proteins such as programmed cell death protein-4 and multiple members of the dedicator of cytokinesis family [56]. To induce miR-21, Smads appear to interact with p68 in the Drosha Microprocessor complex and promote cleavage of pri-miR-21 into pre-miR-21 [34]. Smads have been shown to associate with Drosha and p68 on pri-miR-21 by immunoprecipitation studies. Moreover, Drosha binding to pri-miR-21 is elevated following TGFβ or BMP4 stimulation [34]. On the other hand, Drosha binding to pri-miRNA is reduced and miRNA processing is decreased when Smads are downregulated by small inhibitor RNA, suggesting that Smads are required for Drosha-mediated miRNA processing [34]. Interestingly, unlike in the canonical Smad signaling pathway, a co-Smad is not essential for induction of miRNA processing [34]. Therefore, TGFβ and BMP signals regulate gene expression via miRNAs independently of Smad4.

More recently, the molecular mechanism by which Smads regulate Drosha-mediated miR-21 processing has been characterized [57]. Following ligand stimulation, a cohort of miRNAs is increased without induction of pri-miRNA levels. To identify eventual common characteristics among these miRNAs, the sequences of mature and precursor miRNAs were analyzed. Interestingly, 17 of 20 miRNAs regulated by TGFβ or BMP4 contained a conserved sequence (5′-CAGAC-3′) toward the center of the mature miRNA region, which was termed the RNA–Smad-binding element (R-SBE) [57]. Mutation of the R-SBE in pri-miRNAs abolishes the induction of miRNAs by BMP4, while mutations outside of the R-SBE sequence do not affect the induction of miRNA processing by BMP4 [57]. Thus, an R-SBE is required for ligand-induced miRNA processing [57] (Fig. 6.3). Moreover, when the R-SBE is mutated, Smad association with the pri-miRNA is reduced, as well as binding of Drosha and DGCR8 to the pri-miRNA [57]. Therefore, the Smad proteins are critical mediators of Drosha-dependent miRNA processing upon TGFβ and BMP signaling, and the R–SBE is a required determinant of specificity in Smad-mediated miRNA processing (Fig. 6.3).

Interestingly, the R–SBE is identical to the consensus sequence for DNA binding by Smads [41]. Partially purified recombinant Smad proteins directly associate with pri–miRNA *in vitro* [57]. These observations indicate that Smads bind both DNA and dsRNA through the same sequence, despite differences in the three-dimensional structures of DNA and dsRNA. In solution, dsRNA has a narrower major groove and a wider minor groove compared to B-form DNA [58]. However, other proteins, including the first described eukaryotic transcription factor, TFIIIA, have been shown to bind both DNA and RNA. Therefore, it is plausible that the amino-terminus DNA-binding domain of the R–Smads may be able to bind the RNA helix of pri–miRNAs as well as the B-form helix of DNA [59].

Recent studies suggest that proper RNA binding by the Drosha Microprocessor components is a critical determinant of Drosha cleavage efficiency, affecting mature miRNA size and stability [60,61]. Thus, if it affects recruitment to the Microprocessor complex, the location of the R–SBE within the hairpin structure of the pri–miRNA may be important for generating a mature miRNA sequence. Indeed, when the R–SBE motif was inserted into a heterologous *Caenorhabditis elegans* miRNA (pri–miR-84) at three different positions of the stem region, locating the R–SBE into the mid-region of *C. elegans* pri–miR-84 was sufficient to confer Smad-mediated processing [57], as well as further processing by Dicer and loading onto the RISC [57]. However, insertion of the R–SBE at either the 3′ or 5′ regions of the mature miRNA was unable to confer Smad-mediated processing, suggesting that not only the sequence, but also the position of the R–SBE within the pri–miRNA, is important for Smad-mediated processing [57]. Consistent with this observation, the R–SBE is predominantly localized to the middle of the mature miRNA in endogenous Smad-processed miRNAs [57]. The role of Smads in the regulation of Drosha-mediated pri–miRNA processing has been further supported and nuanced by the study of a nuclear factor called Smad nuclear interacting protein 1 (SNIP1) [62]. SNIP1, originally identified as a nuclear protein partner of Smads, was shown to associate with Drosha and modulate miRNA biogenesis. Downregulation of SNIP1 reduces the expression of a subset of miRNAs, including miR-21 [62]. Therefore, it is possible that Smads alternatively regulate miRNA biogenesis by modulating the association of SNIP1 with Drosha.

6. CLOSING REMARK

In this chapter, we summarized the emerging field of the regulation of miRNA processing at the level of the Drosha complex, focusing in particular on the role of Smad proteins. Increasing evidence suggests that each step of the miRNA biogenesis pathway can be controlled by various extracellular stimuli, including growth factor signaling, to fine-tune cellular miRNA levels. The future challenge is understanding how multiple miRNA regulatory mechanisms coordinate miRNA expression to achieve a precise gene expression landscape under physiological and pathological conditions. The goal, besides unraveling a beautiful biological process, is to help developing novel approaches to facilitate, inhibit, or modulate the activities of pleiotropic growth factors, such as TGFβ and BMPs.

ACKNOWLEDGMENTS

We apologize to colleagues whose references we have not had the opportunity to discuss here due to limitation of space and focus of the chapter. We thank all members of the Hata lab for helpful discussion, especially G. Lagna for critical reading and editing of the chapter. This work is supported by the University of Incheon International Cooperative Research Grant (to H. K.) and NHLBI (HL093154), LeDucq Foundation, and American Heart Association (to A. H.).

REFERENCES

[1] Rodriguez A, et al. Identification of mammalian microRNA host genes and transcription units. Genome Res 2004;14:1902–10.
[2] Saini HK, et al. Genomic analysis of human microRNA transcripts. Proc Natl Acad Sci USA 2007;104:17719–24.
[3] Borchert GM, et al. RNA polymerase III transcribes human microRNAs. Nat Struct Mol Biol 2006;13:1097–101.
[4] Ozsolak F, et al. Chromatin structure analyses identify miRNA promoters. Genes Dev 2008;22:3172–83.
[5] Cai X, et al. Human microRNAs are processed from capped, polyadenylated transcripts that can also function as mRNAs. RNA 2004;10:1957–66.
[6] Zeng Y, Cullen BR. Efficient processing of primary microRNA hairpins by Drosha requires flanking nonstructured RNA sequences. J Biol Chem 2005;280:27595–603.
[7] Han J, et al. Molecular basis for the recognition of primary microRNAs by the Drosha-DGCR8 complex. Cell 2006;125:887–901.
[8] Denli AM, et al. Processing of primary microRNAs by the Microprocessor complex. Nature 2004;432:231–5.
[9] Gregory RI, et al. The Microprocessor complex mediates the genesis of microRNAs. Nature 2004;432:235–40.
[10] Han J, et al. The Drosha-DGCR8 complex in primary microRNA processing. Genes Dev 2004;18:3016–27.

[11] Lee Y, et al. The nuclear RNase III Drosha initiates microRNA processing. Nature 2003;425:415–9.

[12] Bernstein E, et al. Role for a bidentate ribonuclease in the initiation step of RNA interference. Nature 2001;409:363–6.

[13] Hutvagner G, et al. A cellular function for the RNA-interference enzyme Dicer in the maturation of the let-7 small temporal RNA. Science 2001;293:834–8.

[14] Schwarz DS, et al. Asymmetry in the assembly of the RNAi enzyme complex. Cell 2003;115:199–208.

[15] Khvorova A, et al. Functional siRNAs and miRNAs exhibit strand bias. Cell 2003;115:209–16.

[16] Krol J, et al. Structural features of microRNA (miRNA) precursors and their relevance to miRNA biogenesis and small interfering RNA/short hairpin RNA design. J Biol Chem 2004;279:42230–9.

[17] Okamura K, et al. The mirtron pathway generates microRNA-class regulatory RNAs in Drosophila. Cell 2007;130:89–100.

[18] Ruby JG, et al. Intronic microRNA precursors that bypass Drosha processing. Nature 2007;448:83–6.

[19] Berezikov E, et al. Mammalian mirtron genes. Mol Cell 2007;28:328–36.

[20] Hongjiang W, et al. Human RNase III is a 160-kDa protein involved in preribosomal RNA processing. J Biol Chem 2000;275:36957–65.

[21] Conrad C, Rauhut R. Ribonuclease III: new sense from nuisance. Int J Biochem Cell Biol 2002;34:116–29.

[22] Zamore PD. Thirty-three years later, a glimpse at the ribonuclease III active site. Mol Cell 2001;8:1158–60.

[23] Filippov V, et al. A novel type of RNase III family proteins in eukaryotes. Gene 2000;245:213–21.

[24] Fortin KR, et al. Mouse ribonuclease III. cDNA structure, expression analysis, and chromosomal location. BMC Genomics 2002;3:26.

[25] Zhang H, et al. Single processing center models for human Dicer and bacterial RNase III. Cell 2004;118:57–68.

[26] Danin-Kreiselman M, et al. RNAse III-mediated degradation of unspliced pre-mRNAs and lariat introns. Mol Cell 2003;11:1279–89.

[27] Zeng Y, et al. Recognition and cleavage of primary microRNA precursors by the nuclear processing enzyme Drosha. EMBO J 2005;24:138–48.

[28] Morlando M, et al. Primary microRNA transcripts are processed co-transcriptionally. Nat Struct Mol Biol 2008;15:902–9.

[29] Shiohama A, et al. Nucleolar localization of DGCR8 and identification of eleven DGCR8-associated proteins. Exp Cell Res 2007;313:4196–207.

[30] Fukuda T, et al. DEAD-box RNA helicase subunits of the Drosha complex are required for processing of rRNA and a subset of microRNAs. Nat Cell Biol 2007;9:604–11.

[31] Suzuki HI, et al. Modulation of microRNA processing by p53. Nature 2009;460:529–33.

[32] Yamagata K, et al. Maturation of microRNA is hormonally regulated by a nuclear receptor. Mol Cell 2009;36:340–7.

[33] Castellano L, et al. The estrogen receptor-alpha-induced microRNA signature regulates itself and its transcriptional response. Proc Natl Acad Sci USA 2009;106:15732–7.

[34] Davis BN, et al. SMAD proteins control DROSHA-mediated microRNA maturation. Nature 2008;454:56–61.

[35] Endoh H, et al. Purification and identification of p68 RNA helicase acting as a transcriptional coactivator specific for the activation function 1 of human estrogen receptor alpha. Mol Cell Biol 1999;19:5363–72.

[36] Massague J. TGF-beta signal transduction. Annu Rev Biochem 1998;67:753–91.

[37] Kingsley DM. The TGF-beta superfamily: new members, new receptors, and new genetic tests of function in different organisms. Genes Dev 1994;8:133–46.

[38] Lutz M, Knaus P. Integration of the TGF-beta pathway into the cellular signalling network. Cell Signal 2002;14:977–88.

[39] Feng XH, Derynck R. Specificity and versatility in tgf-beta signaling through Smads. Annu Rev Cell Dev Biol 2005;21:659–93.

[40] Heldin CH, et al. TGF-beta signalling from cell membrane to nucleus through SMAD proteins. Nature 1997;390:465–71.

[41] Massague J, et al. Smad transcription factors. Genes Dev 2005;19:2783–810.

[42] Corcoran DL, et al. Features of mammalian microRNA promoters emerge from polymerase II chromatin immunoprecipitation data. PLoS One 2009;4:e5279.

[43] Kato M, et al. TGF-beta activates Akt kinase through a microRNA-dependent amplifying circuit targeting PTEN. Nat Cell Biol 2009;11:881–9.

[44] Chung AC, et al. miR-192 mediates TGF-beta/Smad3-driven renal fibrosis. J Am Soc Nephrol 2010;21:1317–25.

[45] Pandit KV, et al. Inhibition and role of let-7d in idiopathic pulmonary fibrosis. Am J Respir Crit Care Med 2010;182:220–9.

[46] Sun Q, et al. Transforming growth factor-beta-regulated miR-24 promotes skeletal muscle differentiation. Nucleic Acids Res 2008;36:2690–9.

[47] Qin W, et al. TGF-beta/Smad3 signaling promotes renal fibrosis by inhibiting miR-29. J Am Soc Nephrol 2011;22:1462–74.

[48] Long X, Miano JM. Transforming growth factor-beta1 (TGF-beta1) utilizes distinct pathways for the transcriptional activation of microRNA 143/145 in human coronary artery smooth muscle cells. J Biol Chem 2011;286:30119–29.

[49] Davis-Dusenbery BN, et al. Down-regulation of Kruppel-like factor-4 (KLF4) by microRNA-143/145 is critical for modulation of vascular smooth muscle cell phenotype by transforming growth factor-beta and bone morphogenetic protein 4. J Biol Chem 2011;286:28097–110.

[50] Lujambio A, et al. A microRNA DNA methylation signature for human cancer metastasis. Proc Natl Acad Sci USA 2008;105:13556–61.

[51] Lujambio A, Esteller M. How epigenetics can explain human metastasis: a new role for microRNAs. Cell Cycle 2009;8:377–82.

[52] Brueckner B, et al. The human let-7a-3 locus contains an epigenetically regulated microRNA gene with oncogenic function. Cancer Res 2007;67:1419–23.

[53] Nasser MW, et al. Down-regulation of micro-RNA-1 (miR-1) in lung cancer. Suppression of tumorigenic property of lung cancer cells and their sensitization to doxorubicin-induced apoptosis by miR-1. J Biol Chem 2008;283:33394–405.

[54] Saito Y, Jones PA. Epigenetic activation of tumor suppressor microRNAs in human cancer cells. Cell Cycle 2006;5:2220–2.

[55] Massague J, Wotton D. Transcriptional control by the TGF-beta/Smad signaling system. EMBO J 2000;19:1745–54.

[56] Kang H, et al. Bone morphogenetic protein 4 promotes vascular smooth muscle contractility by activating microRNA-21 (miR-21), which down-regulates expression of family of dedicator of cytokinesis (DOCK) proteins. J Biol Chem 2012;287:3976–86.

[57] Davis BN, et al. Smad proteins bind a conserved RNA sequence to promote microRNA maturation by Drosha. Mol Cell 2010;39:373–84.

[58] Shi Y, et al. Crystal structure of a Smad MH1 domain bound to DNA: insights on DNA binding in TGF-beta signaling. Cell 1998;94:585–94.

[59] Theunissen O, et al. RNA and DNA binding zinc fingers in Xenopus TFIIIA. Cell 1995;71:679–90.

[60] Guil S, Caceres JF. The multifunctional RNA-binding protein hnRNP A1 is required for processing of miR-18a. Nat Struct Mol Biol 2007;14:591–6.

[61] Trabucchi M, et al. The RNA-binding protein KSRP promotes the biogenesis of a sub-set of microRNAs. Nature 2009;459:1010–4.

[62] Yu B, et al. The FHA domain proteins DAWDLE in Arabidopsis and SNIP1 in humans act in small RNA biogenesis. Proc Natl Acad Sci USA 2008;105:10073–8.

CHAPTER SEVEN

PIWI Proteins and Their Slicer Activity in piRNA Biogenesis and Transposon Silencing

Kaoru Sato, Haruhiko Siomi[1]

Department of Molecular Biology, Keio University School of Medicine, 35 Shinanomachi, Shinjuku-ku, Tokyo, Japan
[1]Corresponding author: e-mail address: awa403@z2.keio.jp

Contents

Abstract

Common to RNAi and related pathways, which are collectively referred to as RNA silencing, is inactivation of cognate RNA targets by small RNA–Argonaute complexes. Small noncoding RNAs of 20–30 nucleotides (nt) in length function as specificity determinants for the repressive activities of Argonaute-containing effector complexes. Argonaute proteins exhibit small RNA-guided RNA endonuclease activity, often referred to as Slicer activity. Animal germline cells express PIWI subclade proteins of the Argonaute superfamily. They bind Piwi-interacting RNAs (piRNAs) to form effector complexes important for germline development and suppress transposon activity to maintain the integrity of the genome in germline cells, both of which are crucial for the success of future generations. In *Drosophila* ovaries, piRNAs are produced via two distinct pathways: the primary processing and ping-pong cycle pathways. In the ping-pong cycle, PIWI proteins engage in a Slicer-dependent amplification loop between sense and antisense transcripts of a transposon, which consumes transposon sense transcripts (mRNAs), thereby silencing transposons posttranscriptionally. Transposon repressive signals or memory, which initiate the ping-pong cycle, are passed from females to their offspring through the germline transmission of PIWI–piRNA complexes. The ping-pong cycle is conserved among animals, indicating that it arose early in evolution as a form of nucleic acid-based immunity to inactivate transposable elements. The primary processing pathway is a

linear Slicer-independent pathway that operates in ovarian somatic cells. piRNAs produced in the primary pathway are exclusively loaded onto Piwi in ovarian somatic cells.
Nucleases involved in the pathway have long been sought; recently, a gene named
Zucchini was shown to encode a nuclease that processes piRNA intermediates into
mature piRNAs.

1. INTRODUCTION

In RNA silencing pathways, small guide RNAs such as small interfering RNAs (siRNAs) and microRNAs (miRNAs) are produced from long,
double-stranded (ds) RNA by the action of a ribonuclease (RNase) III enzyme called Dicer. They are subsequently loaded onto the RNA-induced
silencing complex (RISC), in which small RNAs serve as the sequence determinants of the silencing pathway by either directing cleavage of homologous mRNA via a small RNA-guided RNA endonuclease or the Slicer
activity of RISC or translational repression and deadenylation coupled with
mRNA decay by RISC [1–3]. The core of RISC is the Argonaute protein,
which directly binds the guide small RNA and exhibits Slicer activity,
defining the protein as a self-contained silencing machine [4–7].

Argonaute proteins were originally defined based on two major protein
motifs: the PAZ and PIWI domains [8]. However, recent structural
studies have revealed that they are typically composed of four conserved domains: N-terminal, PAZ, MID, and PIWI [5,9,10]. The PAZ domain
resembles an OB (oligonucleotide binding) fold domain and recognizes
the 3' hydroxyl end of the small RNA that guides protein target
specificity, whereas the MID domain contacts the phosphate group of the
5' end of the small RNA [9–13]. The PIWI domain adopts a folded
structure similar to that of RNase H enzymes and exhibits Slicer activity
[4,5,7], although a subset of Argonaute proteins do not possess Slicer
activity. Slicer activity is biochemically defined as the Mg^{2+}-dependent
RNA-guided RNA endonuclease activity that directs cleavage of its
cognate mRNA target across from nucleotides 10 and 11, measured
from the 5' end of the small RNA guide strand (Fig. 7.1). That is, Slicer
catalyzes cleavage of the scissile phosphate of the target RNA that lies
between nucleotides 11 and 12, and the resultant cleaved product has a 5'
monophosphate (Fig. 7.1) [14,15]. Argonaute acts as a multi-turnover
enzyme that catalyzes multiple rounds of RNA cleavage [16]. RNase H
enzymes contain three highly conserved catalytic carboxylates, which
comprise the DDE (Asp-Asp-Glu) motif [17]. Argonautes with Slicer

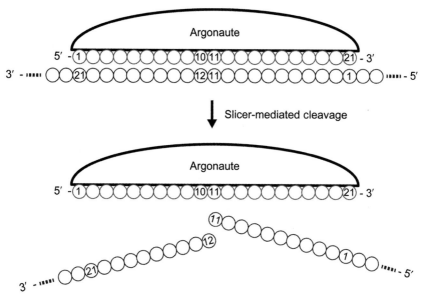

Figure 7.1 Slicer-mediated cleavage of a target transcript. Several Argonaute proteins provide the "Slicer" activity for RNA silencing and catalyze endonucleolytic cleavage, which typically occurs between positions 10 and 11 on the target transcript. (For color version of this figure, the reader is referred to the online version of this chapter.)

activity contain a conserved DDX catalytic triad (where "X" is generally Asp or His) [6,10]. The structure of the Slicer triad, catalytic magnesium ions, and the target RNA strand in the ternary complex, consisting of *Thermus thermophilus* Argonaute, a DNA guide and a 20-mer RNA target, superimposes closely with that of an RNase H catalytic complex [18]. Thus, Slicer employs RNase H-like chemistry to execute slicing of the target RNA strand.

Based on phylogenetic analysis, Argonaute proteins are divided into three subclades: the AGO subclade based on their similarity to its founding member *Arabidopsis* Argonaute 1 (Ago1), which is expressed ubiquitously, the PIWI subclade based on *Drosophila* Piwi (P-element–induced *wi*mpy testis) (Note that hereafter capitalized PIWI stands for either the PIWI subclade proteins or the PIWI domain; Piwi stands for the *Drosophila* protein; italicized *piwi* stands for the gene.), an animal-specific clade that is expressed almost exclusively in gonadal tissues, and the WAGO subclade consisting entirely of *Caenorhabditis elegans*-specific proteins [6,19]. Nearly all eukaryotes, with the conspicuous exception of fission yeast (Baker's yeast), have one or more Argonautes [6,19]. The number of Argonaute genes varies among species, ranging from one in *Schizosaccharomyces pombe* to five in *Drosophila* (two AGOs and three PIWIs) to eight in humans (four AGOs and four PIWIs) to 27 in *C. elegans* (five AGOs, three PIWIs, and 19 WAGOs) [2,20]. Here, we focus on the recent work on the PIWI clade proteins of the model animal *Drosophila melanogaster,* one that has contributed significantly to our understanding of how PIWI subclade proteins mediate transposon silencing. For reviews of PIWI proteins in other animals, see Refs. [21–23].

2. DROSOPHILA PIWI PROTEINS

2.1. Piwi

The founding member of the PIWI subclade is the *Drosophila* Piwi, which was originally discovered owing to its function in germline stem cell (GSC) self-renewal and germline development. Both male and female *Piwi* mutant flies fail to maintain GSCs, and thus they are sterile [24]. Piwi is expressed in both GSCs and surrounding somatic cells and accumulates in the nuclei in both cell types in ovaries [24]. Genetic studies have revealed that expression of Piwi in the surrounding somatic niche cells is required for GSC maintenance, whereas expression of Piwi in GSCs promotes their division [24,25]. In contrast, in testes, Piwi is localized only in the nuclei of somatic cells termed hub cells, which form a niche for GSCs [26]. How does Piwi function in GSC self-renewal and maintenance? A clue to this came from the finding that several transposons were derepressed in *piwi* mutant ovaries and testes, suggesting that Piwi is required for transposon silencing in both male and female gonads [27–33]. Transposable elements move into new sites of the genome either by a copy-and-paste mechanism or by a cut-and-paste mechanism; therefore, they are natural mutagens [21,34]. The GSC genome is vulnerable to chromosomal breaks owing to either meiotic recombination or transposon mobilization [35,36]. Thus, it is crucial to keep transposons silenced to maintain germline genome integrity. How, then, does Piwi suppress transposons? Piwi was found to bind germline-specific small RNAs of 23–29 nt in length, collectively referred to as Piwi-interacting RNAs (piRNAs) (Fig. 7.2) [1,2,21,22], which are longer than siRNAs and miRNAs. The majority of Piwi-associated piRNAs are derived from the antisense strands of various transposons with respect to transposon mRNAs (transposon sense transcripts) [27,28,32,37,38]. Piwi-bound piRNAs carry a 5′ monophosphate group and exhibit a preference for a 5′ uridine (U) residue [27,28,37,39]. Unlike mammalian miRNAs,

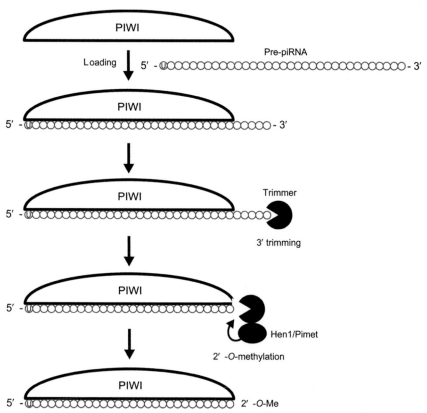

Figure 7.2 piRISC formation. piRNA precursors (pre-piRNAs) can be generated by processing of long transcripts from piRNA clusters (the primary pathway) or cleaved target transcripts mediated by the mature piRISC (the ping-pong cycle). PIWI proteins incorporate long pre-piRNAs with a faithful 5'-nucleotide preference. For example, pre-piRNAs with a 5' monophosphate are loaded onto Siwi, a silkworm Piwi protein, with a strong bias for uridine at position 1 (1U). Then, the loaded pre-piRNAs are trimmed to approximately 27–30 nt by unidentified trimmer(s), and finally 2'-O-methylated at their 3' ends by Hen1/Pimet. 2'-O-methylation is, most likely, coupled with trimming. (For color version of this figure, the reader is referred to the online version of this chapter.)

but similar to plant miRNAs and animal endogenous siRNAs, piRNAs carry a 2'-O-methyl (2'-O-Me) modification at their 3' ends [32,40–44]. This methylation is mediated by HEN1 RNA methyltransferase which likely acts on the piRNA precursor while it is already bound to Piwi [45,46]. The 3' end methylation probably protects piRNAs from the nontemplated addition of uracil residues that in turn trigger degradation of piRNAs [47]. These findings raised the possibility that piRNAs guide the repressive activities of Piwi toward transposon transcripts or transposon DNAs, probably by means of base pairing to repress transposons [27,28,32,48]. The mechanism(s) by which Piwi represses transposons remains to be elucidated; however, mutation analyses have revealed that nuclear localization of Piwi is required for transposon silencing, even though Piwi's Slicer activity appears to be unnecessary for this function [48]. The results of previous studies suggested that Piwi may play a role in heterochromatin formation in transposon loci in the genome [33,49].

2.2. Aub and AGO3

In addition to Piwi, *Drosophila* gonads express two other PIWI proteins: Aubergine (Aub) and AGO3. The *aub* gene was identified as one of the *spindle*-class genes required for microtubule polarizing events that establish the two main body axes, the anterior/posterior (A/P) and dorsal/ventral (D/V) axes, during oogenesis [50,51]. Polarized microtubule cytoskeletons direct the asymmetric localization of RNA molecules including *bicoid, osker*, and *gurken* mRNAs within the oocyte, which is essential for establishing the body axis [52–56]. *Spindle* mutations disrupt each of the symmetry-breaking steps during oogenesis; therefore, they are sterile [57–59]. Recent studies have shown that a number of *spindle*-class genes, including *aub, spindle-E, maelstrom, vasa, armitage, zucchini*, and *squash*, are involved in both piRNA production (see below) and silencing transposons in the germline (Table 7.1) [60–67]. It has been hypothesized that developmental defects observed in *spindle*-class mutants are merely secondary to transposon derepression and mobilization, which activate the DNA damage signaling kinase Chk2, which in turn represses microtubule polarization [35,68]. Aub is expressed only in GSCs and their developing cells in both ovaries and testes, and it accumulates at the nuage, an electron-dense perinuclear organelle [69,70]. Many *spindle*-class proteins were also accumulated in the nuage; thus, it is hypothesized that the nuage is the location of piRNA production and transposon silencing (see below) in the cytoplasm.

Although Aub in the ovary appears to be mainly engaged in transposon silencing, in the testis, Aub is also required for *Stellate* silencing [71–73]. *Stellate* encodes a protein homologous to the regulatory subunit of the protein kinase casein kinase 2 (CK2) [71,74,75]. The X-linked *Stellate* cluster of tandem repeats, which is composed of ~200 copies of the protein-coding gene *Stellate*, is silenced by the ~50 copies of the homologous *Suppressor of Stellate* [*Su(Ste)*] locus on the Y chromosome [71]. Derepression of *Stellate*

Table 7.1 *Drosophila* piRNA genes required for piRNA production in ovary

Gene name	Symbols	CG number	Domains	Mammalian homolog	Notes				Cellular localization		Interaction with PIWIs	Other binding partners[b]	References
					Spindle-class	Suppression of spindle defects[a]			Germ[c]	Germline-soma[d]			
						mei-41	mnk	mei-W68					
piwi	*piwi*	CG6122	PAZ, MID, PIWI	PIWIL 1–4				No	Nucleus	Nucleus	None	Yb, Armi, Zuc	[35,116]
aubergine	*aub*	CG6137	PAZ, MID, PIWI	PIWIL 1–4	Classical	Yes	Yes		Nuage	—	Ago3	Tud, Vas, Spn-E, Tej	[35,116]
argonaute 3	*ago3*	CG40300	PAZ, MID, PIWI	PIWIL 1–4					Nuage	—	Aub	Tud	[116]
partner of piwis	*papi*	CG7082	TUDOR, 2 × KH	TDRD2					Nuage	—	Piwi, Ago3	Tral, Me31B, TER94	[97]
qin (kumo)	*qin (kumo)*	CG14303, CG14306	5 × TUDOR, RING, B-box	TDRD4					Nuage, nucleus	—	Piwi, Aub	Vas, Spn-E, HP1	[103,104]
tejas	*tej*	CG8589	TUDOR, Lotus (Tejas)	TDRD5					Nuage	—	Aub	Vas, Spn-E	[96]
tudor	*tud*	CG9450	11 × TUDOR	TDRD6					Nuage	Cytoplasm	Aub, Ago3	N.D.	[88,116]
spindle-E	*spn-E*	CG3158	TUDOR, DEXDc, HELICc, HA2	TDRD9	Classical	No	No		Nuage	—	Aub	Vas, Tej	[35,96]
FS(1)Yb	*Yb*	CG2706	TUDOR, Helicase-like, DEAD, Helicase, ZF	TDRD12					—	Yb body	Piwi	Armi, Vret	[48,102,108]

Continued

Table 7.1 *Drosophila* piRNA genes required for piRNA production in ovary—cont'd

Gene name	Symbols	CG number	Domains	Mammalian homolog	Spindle-class Notes — Suppression of spindle defects[a]			Cellular localization		Interaction with PIWIs	Other binding partners[b]	References
					mei-41	mnk	mei-W68	Germ[c]	Germline-soma[d]			
Brother of Yb	*BoYb*	CG11133	TUDOR, DEAD, Helicase, ZF	TDRD12				Nuage	–	N.D.	N.D.	[99]
Sister of Yb	*SoYb*	CG31755	2 × TUDOR, DEAD, Helicase, ZF	TDRD12				Cytoplasm	Yb body	N.D.	N.D.	[99]
Krimper	*Krimp*	CG15707	TUDOR	None	No		No	Nuage	N.D.	N.D.	N.D.	[116,124]
vreteno	*vret*	CG4771	2 × TUDOR	None				Nuage	Yb body	Piwi, Aub, Ago3	Yb, Armi	[98]
dSETDB1	*dSETDB1*	CG12196	2 × TUDOR, MBD, PreSET, 2 × SET	SETDB1				Nucleus	Nucleus	N.D.	N.D.	[100]
rhino	*rhi*	CG10683	CHROMO, CHROMO shadow	HP1d	Yes	Yes	No	Nucleus	–	N.D.	Cuff	[117]
cutoff	*cuff*	CG13190	Ral 1-like	DOM3Z	No	Yes	No	Nuage (3 × HA-tagged) nucleus (EGFP-tagged)	–	N.D.	Rhi	[118,119]

vasa	*vas*	CG43081	DEADc, HELICc	VASA	Classical	Yes	Yes	No	Nuage	–	Aub	Tej, Spn-E	[64, 65, 69, 96]
armitage	*armi*	CG11513	Helicase	MOV 10	Classical	Yes	Yes	No	Cytoplasm	Yb body	Piwi	Yb, Vret	[28,35,48,66,102]
zuchini	*zuc*	CG12314	Phospholipase-D/nuclease family	mitoPLD	Classical	Yes	Yes	Yes	Nuage	Mitochondrion	Aub	N.D.	[28,48,66,102,112]
maelstrom	*mael*	CG11254	HMG-box	MAELSTROM	Classical	No	No		Nuage, nucleus, spindle MT	Cytoplasm, nucleus, spindle MT	None	MTOC proteins	[62,69,120]
squash	*squ*	CG4711	Similar to Rnase HII	None	Classical	Yes	No		Nuage	N.D.	Aub	N.D.	[28,48,66]
heat shock protein 83	*hsp83*	CG1242	HSP90, ATPase	HSP90					Any subcellular fraction	Any subcellular fraction	Piwi	HOP	[121,122]
shutdown	*shu*	CG4735	PPlase, TPR	FKBP6					Nuage	Yb body	Piwi	Armi	[123]

[a]Mutations in classical *spindle*-class genes disrupt embryonic axis specification, triggering defects in microtubule organization, localization of mRNAs such as those for *osk* and *grk*, and inefficient *grk* translation in oocytes during oogenesis. Mutations in several piRNA genes also result in similar defects. These defects are dramatically suppressed by mutations in *mei-41* and *mnk*, which encode ATR and Chk2 kinases that function in DNA damage signal transduction, respectively, but not by mutations in *mei-W68*, indicating that the piRNA pathway appears to suppress DNA damage in the *Drosophila* germline, and that mutations in this pathway block axis specification by activation of an ATR/Chk2-dependent DNA damage response that disrupts microtubule polarization, mRNA localization and *grk* translation.

[b]Nurse cells.

[c]Follicle cells or ovarian somatic cells.

[d]Only proteins relating to the piRNA pathway or other RNA silencing mechanisms. B-box, zinc-finger B-box domain; CHROMO, CHRomatin Organization MOdifier domain; CHROMO shadow, a variant of CHROMO; DEAD, DEAD-box helicase; DEXDc, DEAD-like helicase domain; FKBP6, FKBP-type peptidyl-prolyl *cis–trans* isomerase domain; HELICc, helicase-superfamily C-terminal domain; HA2, helicase-associated domain; KH, K homology domain; MBD, methyl-CpG-binding domain; MID, middle; MYND, zinc-finger myeloid-nervy-DEAF-1 domain; PAZ, PIWI/Argonaute/Zwille; PIWI, P-element-induced wimpy testes; PreSET, N-terminal domain to SET; RING, Really Interesting New Gene finger domain; SAM, S-adenosylmethionine; SET, Su(var)3-9/Enhancer-of-zeste/Trithorax domain; SNase, *staphylococcal* nuclease; TUDOR, TUDOR domain; UBA, ubiquitin-associated domain. "—" indicates undetected. N.D. indicates not determined.

occurs in *aub* mutants, resulting in the formation of Stellate protein crystals in primary spermatocytes, which in turn leads to partial or complete male sterility [32,71,72,75,76]. Aub is required for accumulation of small RNAs derived from *Su(Ste)* antisense transcripts. These small RNAs are currently referred to as *Su(Ste)* piRNAs [32,71,73], and are the most abundant class of Aub-associated piRNAs in testes [73,77]. Immunopurified Aub–piRNA complexes from testes display activity in cleaving target RNA-containing sequences complementary to *Stellate* transcripts [73]. These findings are consistent with data demonstrating that mutations in *aub* cause an increase in *Stellate* mRNA and suggest that *Su(Ste)* piRNAs guide the Slicer activity of Aub to *Stellate* transcripts in the cytoplasm to cleave and thereby repress the expression of *Stellate* [73,77]. These findings also indicate that piRNAs can trigger RNA silencing *in trans* and that they also have the potential to regulate the expression of cellular genes.

The second largest class of piRNAs associated with Aub in the testes is derived from a repetitive region on chromosome X, termed AT-chX [73,77]. One of these piRNAs, termed AT-chX-1, shows strong complementarity to *vasa* (*vas*) mRNA, a germline-specific transcript involved in oocyte differentiation and cyst development [63,64]. A target RNA containing a sequence of *vas* mRNA showing complementarity to AT-chX-1 is cleaved by Aub–piRNA complexes immunopurified from testes, consistent with the fact that the protein levels of VAS are increased in *aub* mutant testes [73,78]. This is the second example of piRNA-regulating cellular genes. piRNAs from the two loci *Su(Ste)* and *AT-chX* do not bind Piwi [73], consistent with the fact that Piwi is not involved in *Stellate* silencing.

Loss of Ago3 in the germline causes dorsal appendage defects, which are often observed in *spindle*-class mutants [78]. In *ago3* mutant testes, GSCs are not properly maintained and, therefore, *ago3* males are semifertile [77,78]. Ago3 is required for both transposon silencing in ovaries and accumulation of both *Su(Ste)* and *AT-chX* piRNAs in testes [77,78]. Stellate protein crystals form in primary spermatocytes in *ago3* testes, but not as abundantly as they do in *aub* testes [77]. This is probably due to the much lower levels of AGO3 protein compared with the levels of Aub protein in testes [77]. The expression patterns of AGO3 in ovaries and testes are very similar to those of Aub; namely, AGO3 is cytoplasmic and accumulates at the nuage in GSCs and their developing cells of both ovaries and testes [77,78]. Therefore, Aub and AGO3 are spatially separated from Piwi, which is nuclear and expressed both in germline cells and the surrounding somatic cells.

3. piRNA BIOGENESIS

piRNAs in flies mainly correspond to various types of transposable elements and other repetitive elements [21,39]. Mutations in fly Dicer do not affect piRNA production, indicating that biogenesis of piRNAs is distinct from that of miRNAs and siRNAs, and does not involve dsRNA precursors [32]. piRNAs are thought to be transcribed from transposon-rich clusters in the genome as long single-stranded precursors that are processed by single-stranded RNA-specific nucleases into mature piRNAs [37]. Mature piRNAs carry a monophosphate at their 5′ ends and a 2′-O-Me modification at their 3′ ends [32,40–46] and guide transposon silencing in *trans*. Sequencing of piRNAs has revealed that two main pathways exist to generate piRNAs in the *Drosophila* ovary, the ping-pong cycle and primary pathways [28,29,37,38,43] (Fig. 7.3).

3.1. The ping-pong cycle pathway of piRNA production

piRNAs associated with Aub and Piwi are derived mainly from the antisense strands of retrotransposons, while AGO3-associated piRNAs arise mainly from the sense strands [21,37,38]. Aub- and Piwi-associated piRNAs show a strong preference for uracil at their 5′ ends, and AGO3-associated piRNAs show a preference for adenine at nucleotide 10, with no 5′ nt preference [21,37,38]. The first 10 nt of Aub-associated piRNAs can be perfectly complementary to the first 10 nt of AGO3-associated piRNAs [21,37,38]. The overlap by precisely 10 nt at their 5′ ends suggests the involvement of the Slicer activity of PIWI proteins in piRNA biogenesis (Figs. 7.1 and 7.4). PIWI proteins retain Slicer activity that allows them to cleave an RNA substrate across from position 10 of their bound piRNA [21,37,38]. Slicer-mediated cleavage determines and produces the 5′ end of piRNAs that are derived from the opposite strand of transposon transcripts [21,37,38]. The Slicer activities of AGO3 and Aub act catalytically, thereby leading to repeated rounds of piRNA production, forming a feed-forward amplification loop (Fig. 7.4). This amplification loop is often referred to as the ping-pong cycle [21,37]. How, then, are the 3′ ends of piRNAs formed? After loading of the resulting cleavage products onto another PIWI protein, a second nuclease activity generates the 3′ end of the piRNA with the specific size determined by the footprint of the particular PIWI member on the RNA, which appears to precede the 2′-O-Me modification [46,79]. In each PIWI protein, the PAZ domain may be positioned at a distance from the MID domain that corresponds to the

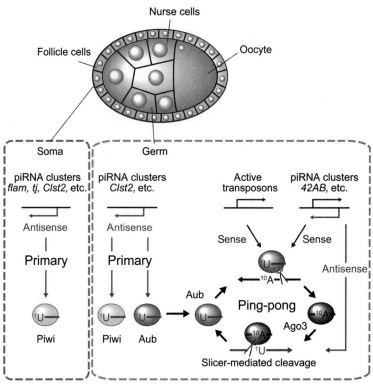

Figure 7.3 Two piRNA biogenesis pathways in the Drosophila ovary. The three PIWI pro-teins—Piwi, Aubergine (Aub), and Argonaute3 (Ago3)—are expressed in *Drosophila* ovarian germ cells (nurse cells and oocytes), whereas only Piwi is expressed in ovarian somas (e.g., follicle cells). In the primary piRNA processing pathway, primary antisense transcripts transcribed from piRNA clusters and/or transposons are processed to piRNAs by unknown mechanisms and are loaded onto Aub or Piwi with 1U. piRNAs derived from the *flam* locus are exclusively loaded onto Piwi because *flam* is active only in ovar-ian somas. In germ cells, the Piwi–piRNA complex or piRNA-induced silencing com-plexes (piRISCs) produced through this mechanism act as a "trigger" of the amplification loop. The amplification loop (also known as the ping-pong cycle) most likely involves the Slicer activities of Aub and Ago3, but not that of Piwi itself. Aub as-sociated with antisense piRNAs cleaves piRNA precursors in the sense strand. This reac-tion determines and forms the 5′ end of piRNAs that are loaded onto Ago3 with a strong bias for adenine at position 10 (10A). Ago3 associated with sense piRNAs cleaves anti-sense piRNA precursors, generating the 5′ end of antisense piRNAs that are subse-quently loaded onto Aub with 1U. (See color plate section in the back of the book.)

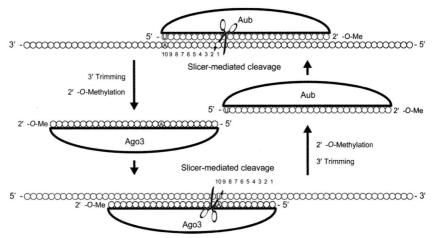

Figure 7.4 The ping-pong amplification loop. Aub associates predominantly with 1U-piRNAs arising from the antisense transcripts of transposons. The Aub–piRNA complex guides Slicer-dependent cleavage of TE transcripts in the sense strand, yielding sense piRNA precursors with 10A. How the 3′ end of piRNAs is formed remains unknown. Reciprocally, Ago3-associated piRNAs guide cleavage of antisense TE transcripts, yielding antisense piRNA precursors with 1U. The 3′ end of piRNA is trimmed by an unknown nuclease (or nucleases), which is followed by 2′-O-methylation mediated by Hen1/Pimet. piRNAs that induce the amplification loop may also be deposited from the mother. (For color version of this figure, the reader is referred to the online version of this chapter.)

length of each piRNA [5,9–13]. It is likely that the second, as yet unidentified nuclease, with a 3′–5′ exonucleolytic activity, trims long cleaved piRNA precursors with their 5′ ends anchored onto PIWI proteins.

These observations suggest the existence of a self-amplifying loop for piRNAs in which antisense piRNAs in Aub direct the cleavage of sense transcripts, transposon mRNAs, and guide the formation of the 5′ end of sense piRNA in AGO3, and vice versa [21,37,38]. In this loop, transposon transcripts are cleaved and consumed by PIWI–piRNA complexes, and therefore, transposons are posttranscriptionally silenced. That is, transposons are both a source gene for piRNAs and a target of piRNA-mediated silencing in the ping-pong cycle [21,37,38]. In other words, transposon loci would be expressed continuously, but their transcripts are cleaved continuously in the cycle. It is also implied that the amplification loop would steer piRNA production toward transcriptionally active transposable elements. The signature of this amplification cycle—a 10-nt overlap at the 5′ ends of sense and antisense piRNAs—is also apparent in many eukaryotes including sea anemones, sponges, silkworms, zebrafish, and mammals [80–85].

The ping-pong cycle mainly engages AGO3 and Aub, both of which are cytoplasmic proteins that accumulate at the nuage in GSCs and their developing cells but not in the surrounding somatic cells [27,37,38,73]. Thus, the cycle operates specifically in germline cells. Nuages are perinuclear structures found at the cytoplasmic face of the nuclear envelope in animal germline cells [69,70,86]. Genetic studies have identified a number of genes required for piRNA production (Table 7.1). These include *spindle-class* genes. Although the precise functions of these genes in the ping-pong cycle remain to be elucidated, the protein products of many of these genes are also enriched in the nuage, suggesting that the ping-pong cycle may operate there. In the ping-pong cycle, the resulting cleavage products of piRNA precursors are loaded onto another PIWI protein. Thus, AGO3 and Aub are believed to be located close to the nuage. These PIWI proteins contain symmetric dimethylarginines (sDMAs), which are recognized and bound by Tudor (Tud) domain-containing proteins [87–95]. Several proteins with Tud domains have been identified to be required for piRNA production and found to be accumulated in the nuage in the ovary [29,70,88,96–104]. These Tud proteins are associated with PIWI proteins through their sDMA modification, and one Tud protein, Tudor, was found to form heteromeric complexes with both Aub and AGO3 [88]. Thus, Tud proteins, and particularly those with multiple Tud domains, are thought to act as platforms or ping-pong tables where AGO3 and Aub are located close enough to hand their cleaved products to each other [88] (Fig. 7.4).

There must also be primary piRNAs that initiate this cycle. It is believed that the ping-pong cycle is primed at least partly by maternally contributed piRNAs. The maternal loading of Piwi proteins into embryos is observed in flies [105]. Thus, the maternal contribution of PIWI proteins, and presumably their associated piRNAs, may imply that the primary piRNAs that initiate the ping-pong cycle of piRNA biogenesis are directly deposited from mothers to offspring through germline transmission within PIWI–piRNA complexes [105]. Thus, the ping-pong cycle may also operate between generations. However, the amplification cycle in flies engages mainly AGO3 and Aub, and Piwi is spatially separated from them at the subcellular and cell-type levels but still associates with piRNAs [21,37,38]. In addition, piRNAs derived from a particular piRNA cluster (e.g., the *flamenco* locus, see below) associate almost exclusively with Piwi. These observations indicate that some other mechanisms of piRNA biogenesis must operate in the ovary.

3.2. The primary processing pathway of piRNA production

Ovarian somatic piRNAs that correspond to transposons are almost exclusively derived from one genomic cluster named the *flamenco* locus, via a pathway independent of the ping-pong cycle, and loaded onto Piwi, the only PIWI expressed in ovarian somatic cells [28,29]. This pathway is currently referred to as the primary processing pathway (Fig. 7.3). Unlike the ping-pong pathway, the somatic piRNA pathway does not require the Slicer activity of Piwi, and ping-pong signatures are not apparent in the somatic piRNA population, indicating that there is no Piwi-only ping-pong to produce somatic piRNAs [28]. Genetic and biochemical analyses have revealed that this pathway requires the RNA helicase Armitage (Armi), the Tudor domain-containing RNA helicase Yb, and the mitochondrial phospholipase D homolog Zucchini (Zuc), as essential factors for primary piRNA biogenesis [30,48,101,102]. Armi was originally identified as a gene required for *osker* mRNA silencing and embryonic axis specification [66]. Both plant and mammalian homologs (SDE3 and Mov10, respectively) of Armi have been implicated in RNA silencing [106,107]. *Yb* was originally identified as an essential gene required for GSC self-renewal, and the protein product was found to be expressed in surrounding somatic cells and accumulated in granules in the cytoplasm of follicle cells in the ovary [108]. These granules, named Yb bodies, often localize in close proximity to mitochondria [108]. Armi forms complexes with Piwi and Yb, and is also localized at Yb bodies [48,101]. Mutations in *armi* or *yb* genes lead to a massive decrease in piRNA levels and transposon derepression [30,48,101,102]. The cytoplasmic localization of these proteins suggests that piRNA biogenesis and loading occur in the cytoplasm (Fig. 7.5). RNAi depletion of Armi or Yb results in the cytoplasmic localization of Piwi not loaded with piRNAs [48]. These findings suggest a model in which piRNA precursors are processed in the cytoplasm, and formation of a functional Piwi–piRNA complex occurs in Yb bodies before nuclear entry to exert transposon silencing. In other words, the Yb body is the cellular location for inspecting the function of piRISCs and allowing only functional complexes to be transported into the nucleus.

Zuc was originally identified as a gene required early during oogenesis for the translational silencing of *osk* mRNA, and in the later stages for the proper expression of the Grk protein [67]. Zuc was also found to be required for the biogenesis of piRNAs in ovaries and testes [67]. *Zuc* encodes a member of the phospholipase D (PLD) superfamily and is highly conserved among

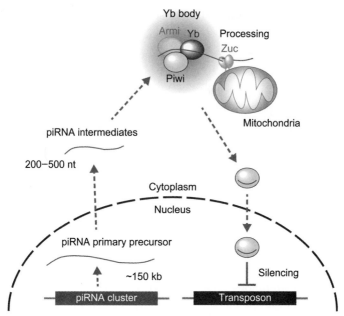

Figure 7.5 The primary piRNA biogenesis in follicle cells of the ovary. In follicle cells, Piwi, but not Aub or Ago3, is expressed. Armi associates with Piwi and causes it to localize to Yb bodies. Yb is the main component of Yb bodies. The piRNA intermediates (intermediate), partially processed from primary piRNA precursors (pre-piRNAs), are loaded onto a complex comprising Armi, Piwi, and Yb at Yb bodies, and are processed into mature piRNAs. Zuc, an endonuclease that functions as a dimer and localizes on the surfaces of mitochondria, is required for primary piRNA processing. Without piRNA loading, PIWI is not localized to the nucleus. The piRISC complex may be associated with other nuclear proteins to form a larger functional complex (a nuclear effector complex) to repress TEs in the nucleus. (See color plate section in the back of the book.)

animal species [109]. Recent studies have shown that Zuc is expressed in ovarian somatic cells and is required for the primary pathway [28–30,48]. In contrast, piRNA production through the ping-pong cycle is only slightly affected in *zuc* mutant ovaries [29]. Zuc contains a mitochondrial localization sequence (MLS) at its N-terminus, and human Zuc is localized on the outer surfaces of mitochondria [109]. Mouse Zuc was also shown to be required for piRNA biogenesis and transposon silencing in testes [110]. However, the ping-pong cycle seems to be still active in the testes of mice with mutant Zuc. Thus, the role of Zuc in piRNA biogenesis appears to be conserved from flies to mammals, and Zuc may play an important role in the primary piRNA processing pathway rather than in the ping-pong cycle.

Fly Zuc was also shown to be located on the outer surfaces of mitochondria [48]. Piwi accumulates at enlarged Yb bodies and is not loaded with piRNAs in Zuc-depleted ovarian somatic cells [48]. Thus, although it is a mitochondrial protein, Zuc regulates the primary piRNA processing pathway in the cytoplasm of ovarian somatic cells. Unlike the classic mammalian PLDs that encode two copies of the highly conserved HKD motif that juxtapose to form the catalytic site, Zuc encodes just a single HKD half-catalytic site [109]. In addition, Zuc shares amino acid sequence similarity with a bacterial nuclease, Nuc, another PLD superfamily member [67,109,111]. Therefore, an important question that needed resolving was whether Zuc is a PLD or a nuclease or both. Very recently, using a combination of structural biology approaches and biochemical experiments, this critical question was answered. The crystal structure of Zuc revealed that the protein forms a dimer and that the HKD motifs in the two protomers (subunits of a dimer) form a single, positively charged, narrow catalytic groove at the dimer interface [112,113]. The active-site groove of Zuc appears too narrow to accommodate PLD substrates such as cardiolipin, because such PLD substrates have bulky hydrophobic lipid tails [112,113]. Indeed, purified Zuc fails to convert cardiolipin to phosphatidic acid *in vitro* [112,113]. Although the structure of Zuc superimposes well with the structure of bacterial Nuc, which can cleave both single- and double-stranded RNAs, the active site of Zuc could accommodate a single-stranded, but not a double-stranded, substrate RNA. Purified Zuc exhibits single-stranded RNA-specific endoribonuclease activity *in vitro* [112,113]. A structural comparison of Zuc with Nuc revealed that the conserved residues are very similarly arranged, suggesting that Zuc cleaves a substrate phosphodiester linkage through a mechanism similar to the proposed two-step catalytic mechanism for Nuc [114,115], in which the HKD motif in one protomer makes a nucleophilic attack on the substrate phosphorus atom and cleaves the phosphodiester bond, while the HKD motif in the other protomer donates a proton to the leaving group of the substrate. The resultant phosphoenzyme intermediate is hydrolyzed by a water molecule that is activated by the donor HKD motif.

Although Zuc shows little sequence specificity to cleave, its cleavage products possess a 5′-monophosphate, a hallmark of mature piRNAs [112]. Mutational analyses have revealed that conserved active-site residues of Zuc are critical for RNase activity *in vitro* and transposon silencing *in vivo* [112]. As mentioned above, depletion of Zuc in ovarian somatic cells causes a severe reduction in the level of mature piRNAs [48]. Interestingly, piRNA intermediate-like molecules of 200–500 nt in length accumulate in Zuc-depleted ovarian somatic cells [112]. The *flamenco* locus is believed to produce a long, ~150 kb single-stranded transcript as a piRNA precursor [37]. Thus, these molecules of 200–500 nt appear to be processing intermediates (Fig. 7.5). Notably, expression of full-length wild-type Zuc resulted in an increase in the level of mature piRNAs and a concomitant decrease in the level of piRNA intermediates [112]. In contrast, expression of mutants that contain amino acid changes in the HKD motif failed to decrease the level of piRNA intermediates in Zuc-depleted ovarian somatic cells. Taken together, these results indicate that Zuc is an endoribonuclease whose activity is required for piRNA maturation [112]. These findings suggest a model for the primary processing pathway of piRNA biogenesis in *Drosophila* ovarian somatic cells in which nascent piRNA-free Piwi is localized to Yb bodies, where it joins the Armi–Yb complex. Next, piRNA intermediates are processed and loaded onto Piwi to form active piRNA–Piwi complexes or piRISCs. Zuc is anchored on the surfaces of mitochondria by its N-terminal MLS as a dimer, leaving the catalytic site facing the cytosol. In this topological environment, Zuc has unrestricted access to its substrates, piRNA intermediates, which are held in the Armi–Yb complex, thereby facilitating primary piRNA production.

4. OUTSTANDING QUESTIONS

piRNA biogenesis pathways and the ways in which piRNA–PIWI complexes repress transposons are complex and diverse. Many questions remain unanswered; some outstanding examples follow:

1. In RNA silencing, one protein, Argonaute, binds to a large number of small RNAs with various sequences, resulting in the formation of effector RISC complexes. Because target recognition uses complementary RNA sequences, once a particular element or gene is recognized by the RNA silencing system, all copies within a cell will be targets for inactivation. The production of small guide RNAs must be carefully controlled to prevent inappropriate silencing. Therefore, this system requires gatekeepers to ensure that only small RNA precursors, but not other cellular RNAs, are recognized by the small RNA biogenesis pathways and that Argonaute can bind small guide RNAs but not degrade cellular small RNAs. How does the ovarian cell distinguish between piRNA precursors or transposon transcripts and transcripts from cellular counterparts, and funnel only the precursors into piRNA biogenesis pathways? How does Piwi

interact with mature piRNAs but not other small RNAs? In other words, how do the piRNA pathways avoid self-directed reactions?

2. While amplification of the silencing signal in the ping-pong cycle would have obvious benefits for suppressing the expression of transposons and repetitive elements, there should be a system that acts to enforce on piRNA pools an antisense bias to efficiently silence transposons by cleaving transposon sense transcripts or mRNAs. Indeed, sense piRNAs are less abundant and disproportionately bind to AGO3. AGO3–sense piRNA complexes act to amplify antisense piRNA pools bound to Aub, increasing the number of piRNAs that can act to destroy transposon mRNAs. Therefore, AGO3, guided by sense piRNAs, lies at the center of the amplification loop. How does AGO3 interact with sense piRNAs? In other words, how does the ping-pong cycle avoid loading antisense piRNAs onto AGO3?

3. Zuc has been identified as a nuclease that converts piRNA intermediates into mature piRNAs. A P-element insertion at the beginning of the *flamenco* locus abrogates piRNA production over the entire ~150-kb cluster, strongly arguing for a long, single-stranded transcript as a piRNA precursor [37]. The long piRNA precursor must be exported to the cytoplasm and processed into intermediates to be further cleaved by Zuc in the cytoplasm. How is such a long transcript exported to the cytoplasm and which nuclease(s) specifically recognizes and cleaves the precursor to produce substrates for Zuc?

4. Although the ping-pong cycle operates in fly testes [77], how piRNAs are produced in fly testes remains largely unknown. A large number of piRNAs with exactly the same sequences, derived from antisense strands of the two loci *Su(Ste)* and *AT-chX*, are associated with both AGO3 and Aub [77]. Because the production of *Su(Ste)* piRNAs, but not *AT-chX* piRNAs, depends on the activity of Armi, these piRNAs seem to be produced by distinct mechanisms. [77] How are these piRNAs produced? How are piRNAs loaded onto Piwi produced in testes?

5. In addition to ovarian somatic cells, primary piRNAs also appear to be produced in germline cells in the ovary and to be loaded onto both Piwi and Aub, which then could initiate the ping-pong cycle. However, neither of Armi and Yb proteins are expressed in germline cells in the ovary. How are primary piRNAs produced in germline cells in the ovary?

6. Piwi represses transposons in the nucleus. Heterochromatin formation appears to be involved in transposon silencing mediated by Piwi–piRNA complexes in the nucleus. How do Piwi–piRNA complexes direct

heterochromatin formation at target transposon loci? Do Piwi–piRNA complexes directly interact with histone modification enzymes that initiate and maintain heterochromatin?

We look forward to the day when these and other important questions are answered.

REFERENCES

[1] Ghildiyal M, Zamore PD. Small silencing RNAs: an expanding universe. Nat Rev Genet 2009;10:94–108.

[2] Kim NV, Han J, Siomi MC. Biogenesis of small RNAs in animals. Nat Rev Mol Cell Biol 2009;10:126–39.

[3] Siomi H, Siomi MC. On the road to reading the RNA-interference code. Nature 2009;457:396–404.

[4] Liu J, Carmell MA, Rivas FV, Marsden CG, Thomson JM, Song JJ, et al. Argonaute2 is the catalytic engine of mammalian RNAi. Science 2004;305:1437–41.

[5] Song JJ, Smith SK, Hannon GJ, Joshua-Tor L. Crystal structure of Argonaute and its implications for RISC slicer activity. Science 2004;305:1434–7.

[6] Joshua-Tor L, Hannon GJ. Ancestral roles of small RNAs: an Ago-centric perspective. Cold Spring Harb Perspect Biol 2011;3:a003772.

[7] Meister G, Landthaler M, Patkaniowska A, Dorsett Y, Teng G, Tuschl T. Human Argonaute2 mediates RNA cleavage targeted by miRNAs and siRNAs. Mol Cell 2004;15:185–97.

[8] Cerutti L, Mian N, Bateman A. Domains in gene silencing and cell differentiation proteins: the novel PAZ domain and redefinition of the Piwi domain. Trends Biochem Sci 2000;25:481–2.

[9] Yuan YR, Pei Y, Ma JB, Kuryavyi V, Zhadina M, Meister G, et al. Crystal structure of *A. aeolicus* Argonaute, a site-specific DNA-guided endoribonuclease, provides insights into RISC-mediated mRNA cleavage. Mol Cell 2005;19:405–19.

[10] Parker JS. How to slice: snapshots of Argonaute in action. Silence 2010;1:3.

[11] Rivas FV, Tolia NH, Song JJ, Aragon JP, Liu J, Hannon GJ, et al. Purified Argonaute2 and an siRNA form recombinant human RISC. Nat Struct Mol Biol 2005;12:340–9.

[12] Elkayam E, Kuhn CD, Tocilj A, Haase AD, Greene EM, Hannon GJ, et al. The structure of human Argonaute-2 in complex with miR-20a. Cell 2012;150:100–10.

[13] Nakanishi K, Weinberg DE, Bartel DP, Patel DJ. Structure of yeast Argonaute with guide RNA. Nature 2012;486:368–74.

[14] Martinez J, Tuschl T. RISC is a 5′ phosphomonoester-producing RNA endonuclease. Genes Dev 2004;18:975–80.

[15] Schwarz DS, Tomari Y, Zamore PD. The RNA-induced silencing complex is a Mg^{2+}-dependent endonuclease. Curr Biol 2004;14:787–91.

[16] Hutvagner G, Zamore PD. A microRNA in a multiple-turnover RNAi enzyme complex. Science 2002;297:2056–60.

[17] Yang W, Steitz TA. Recombining the structures of HIV integrase, RuvC and RNase H. Structure 1995;3:131–4.

[18] Wang Y, Juranek S, Li H, Sheng G, Tuschl T, et al. Structure of an argonaute silencing complex with a seed-containing guide DNA and target RNA duplex. Nature 2008;456:921–6.

[19] Tolia NH, Joshua-Tor L. Slicer and the argonautes. Nat Chem Biol 2007;3:36–43.

[20] Yigit E, Batista PJ, Bei Y, Pang KM, Chen CC, Tolia NH, et al. Analysis of the C. elegans Argonaute family reveals that distinct Argonautes act sequentially during RNAi. Cell 2006;127:747–57.

[21] Siomi MC, Sato K, Pezic C, Aravin AA. PIWI-interacting small RNAs: the vanguard of genome defence. Nat Rev Mol Cell Biol 2011;12:246–58.

[22] Saito K, Siomi MC. Small RNA-mediated quiescence of transposable elements in animals. Dev Cell 2010;19:687–97.

[23] Pillai RS, Chuma S. piRNAs and their involvement in male germline development in mice. Dev Growth Differ 2012;54:78–92.

[24] Cox DN, Chao A, Baker J, Chang L, Qiao D, Lin H. A novel class of evolutionarily conserved genes defined by *piwi* are essential for stem cell self-renewal. Genes Dev 1998;12:3715–27.

[25] Szakmary A, Cox DN, Wang Z, Lin H. Regulatory relationship among *piwi*, *pumilio*, and *bag-of-marbles* in *Drosophila* germline stem cell self-renewal and differentiation. Curr Biol 2005;15:171–8.

[26] Cox DN, Chao A, Lin H. piwi encodes a nucleoplasmic factor whose activity modulates the number and division rate of germline stem cells. Development 2000;127:503–14.

[27] Saito K, Nishida KM, Mori T, Kawamura Y, Miyoshi K, Nagami T, et al. Specific association of Piwi with rasiRNAs derived from retrotransposon and heterochromatic regions in the *Drosophila* genome. Genes Dev 2006;20:2214–22.

[28] Saito K, Inagaki S, Mituyama T, Kawamura Y, Ono Y, Sakota E, et al. A regulatory circuit for piwi by the large Maf gene traffic jam in *Drosophila*. Nature 2009;461:1296–301.

[29] Malone CD, Brennecke J, Dus M, Stark A, McCombie WR, Sachidanandam R, et al. Specialized piRNA pathways act in germline and somatic tissues of the *Drosophila* ovary. Cell 2009;137:522–35.

[30] Haase AD, Fenoglio S, Muerdter F, Guzzardo PM, Czech B, Pappin DJ, et al. Probing the initiation and effector phases of the somatic piRNA pathway in *Drosophila*. Genes Dev 2010;24:2499–504.

[31] Zamparini AL, Davis MY, Malone CD, Vieira E, Zavadil J, Sachidanandam R, et al. Vreteno, a gonad-specific protein, is essential for germline development and primary piRNA biogenesis in *Drosophila*. Development 2011;138:4039–50.

[32] Vagin VV, Sigova A, Li C, Seitz H, Gvozdev V, Zamore PD. A distinct small RNA pathway silences selfish genetic elements in the germline. Science 2006;313:320–4.

[33] Klenov MS, Lavrov SA, Stolyarenko AD, Ryazansky SS, Aravin AA, Tuschl T, et al. Repeat-associated siRNAs cause chromatin silencing of retrotransposons in the *Drosophila melanogaster* germline. Nucleic Acids Res 2007;35:5430–8.

[34] Kazazian Jr. HH. Mobile elements: drivers of genome evolution. Science 2004;303:1626–32.

[35] Klattenhoff C, Bratu DP, McGinnis-Schultz N, Koppetsch BS, Cook HA, Theurkauf WE. *Drosophila* rasiRNA pathway mutations disrupt embryonic axis specification through activation of an ATR/Chk2 DNA damage response. Dev Cell 2007;12:45–55.

[36] Theurkauf WE, Klattenhoff C, Bratu DP, McGinnis-Schultz N, Koppetsch BS, Cook HA. rasiRNAs, DNA damage, and embryonic axis specification. Cold Spring Harb Symp Quant Biol 2006;71:171–80.

[37] Brennecke J, Aravin AA, Stark A, Dus M, Kellis M, Sachidanandam R, et al. Discrete small RNA-generating loci as master regulators of transposon activity in *Drosophila*. Cell 2007;128:1089–103.

[38] Gunawardane LS, Saito K, Nishida KM, Miyoshi K, Kawamura Y, Nagami T, et al. A Slicer-mediated mechanism for repeat-associated siRNA 5′ end formation in *Drosophila*. Science 2007;315:1587–90.

[39] Aravin AA, Lagos-Quintana M, Yalcin A, Zavolan M, Marks D, Snyder B, et al. The small RNA profile during *Drosophila melanogaster* development. Dev Cell 2003;5:337–50.

[40] Pelisson A, Sarot E, Payen-Groschene G, Bucheton A. A novel repeat-associated small interfering RNA-mediated silencing pathway downregulates complementary sense gypsy transcripts in somatic cells of the *Drosophila* ovary. J Virol 2007;81:1951–60.

[41] Kirino Y, Mourelatos Z. The mouse homolog of HEN1 is a potential methylase for Piwi-interacting RNAs. RNA 2007;13:1397–401.

[42] Ohara T, Sakaguchi Y, Suzuki T, Ueda H, Miyauchi K, Suzuki T. The 3' termini of mouse Piwi-interacting RNAs are 2'-O-methylated. Nat Struct Mol Biol 2007;14:349–50.

[43] Li J, Yang Z, Yu B, Liu J, Chen X. Methylation protects miRNAs and siRNAs from a 3'-end uridylation activity in Arabidopsis. Curr Biol 2005;15:1501–7.

[44] Yu B, Yang Z, Li J, Minakhina S, Yang M, Padgett RW, et al. Methylation as a crucial step in plant microRNA biogenesis. Science 2005;307:932–5.

[45] Horwich MD, Li C, Matranga C, Vagin V, Farley G, Wang P, et al. The *Drosophila* RNA methyltransferase, DmHen1, modifies germline piRNAs and single-stranded siRNAs in RISC. Curr Biol 2007;17:1265–72.

[46] Saito K, Sakaguchi Y, Suzuki T, Suzuki T, Siomi H, Siomi MC. Pimet, the *Drosophila* homolog of HEN1, mediates 2'-O-methylation of Piwi- interacting RNAs at their 3' ends. Genes Dev 2007;21:1603–8.

[47] Ameres SL, Horwich MD, Hung JH, Xu J, Ghildiyal M, Weng Z, et al. Target RNA-directed trimming and tailing of small silencing RNAs. Science 2010;328:1534–9.

[48] Saito K, Ishizu H, Komai M, Kotani H, Kawamura Y, Nishida KM, et al. Roles for the Yb body components Armitage and Yb in primary piRNA biogenesis in *Drosophila*. Genes Dev 2010;24:2493–8.

[49] Pal-Bhadra M, Leibovitch BA, Gandhi SG, Rao M, Bhadra U, Birchler JA, et al. Heterochromatic silencing and HP1 localization in *Drosophila* are dependent on the RNAi machinery. Science 2004;303:669–72.

[50] Schmidt A, Palumbo G, Bozzetti MP, Tritto P, Pimpinelli S, Schäfer U, et al. Genetic and molecular characterization of sting, a gene involved in crystal formation and meiotic drive in the male germ line of Drosophila melanogaster. Genetics 1999;151:749–60.

[51] Wilson JE, Connell JE, Macdonald PM. aubergine enhances oskar translation in the *Drosophila* ovary. Development 1996;122:1631–9.

[52] Neuman-Silberberg FS, Schupbach T. The *Drosophila* dorsoventral patterning gene gurken produces a dorsally localized RNA and encodes a TGF α-like protein. Cell 1993;75:165–74.

[53] Neuman-Silberberg FS, Schupbach T. The *Drosophila* TGFα-like protein Gurken: expression and cellular localization during *Drosophila* oogenesis. Mech Dev 1996;59:105–13.

[54] Brendza RP, Serbus LR, Duffy JB, Saxton WM. A function for kinesin I in the posterior transport of oskar mRNA and Staufen protein. Science 2000;289:2120–2.

[55] Schnorrer F, Bohmann K, Nusslein-Volhard C. The molecular motor dynein is involved in targeting swallow and bicoid RNA to the anterior pole of *Drosophila* oocytes. Nat Cell Biol 2000;2:185–90.

[56] Arn EA, Cha BJ, Theurkauf WE, Macdonald PM. Recognition of a bicoid mRNA localization signal by a protein complex containing Swallow, Nod, and RNA binding proteins. Dev Cell 2003;4:41–51.

[57] Tearle R, Nüsslein-Volhard C. Tübingen mutants stocklist. Drosoph Inf Serv 1987;66:209–26.

[58] Schupbach T, Wieschaus E. Female sterile mutations on the second chromosome of *Drosophila melanogaster*. II. Mutations blocking oogenesis or altering egg morphology. Genetics 1991;129:1119–36.

[59] Gonzalez-Reyes A, Elliott H, St Johnston D. Oocyte determination and the origin of polarity in *Drosophila*: the role of the spindle genes. Development 1997;124:4927–37.

[60] Gillespie DE, Berg CA. homeless is required for RNA localization in *Drosophila* oogenesis and encodes a new member of the DE-H family of RNA-dependent ATPases. Genes Dev 1995;9:2495–508.

[61] Wilson JE, Connell JE, Macdonald PM. *aubergine* enhances oskar translation in the *Drosophila* ovary. Development 1996;122:1631–9.

[62] Clegg NJ, Frost DM, Larkin MK, Subrahmanyan L, Bryant Z, Ruohola-Baker H. *maelstrom* is required for an early step in the establishment of *Drosophila* oocyte polarity: posterior localization of *grk* mRNA. Development 1997;124:4661–71.

[63] Lasko PF, Ashburner M. The product of the *Drosophila* gene vasa is very similar to eukaryotic initiation factor-4A. Nature 1988;335:611–7.

[64] Styhler S, Nakamura A, Swan A, Suter B, Lasko P. *vasa* is required for GURKEN accumulation in the oocyte, and is involved in oocyte differentiation and germline cyst development. Development 1998;125:1569–78.

[65] Tomancak P, Guichet A, Zavorszky P, Ephrussi A. Oocyte polarity depends on regulation of gurken by Vasa. Development 1998;125:1723–32.

[66] Cook HH, Koppetsch BS, Wu J, Theurkauf WE. The *Drosophila* SDE3 homolog armitage is required for oskar mRNA silencing and embryonic axis specification. Cell 2004;116:817–29.

[67] Pane A, Wehr K, Schüpbach T. zucchini and squash encode two putative nucleases required for rasiRNA production in the *Drosophila* germline. Dev Cell 2007;12:851–62.

[68] Khurana JS, Theurkauf W. piRNAs, transposon silencing, and Drosophila germline development. J Cell Biol 2010;191:905–13.

[69] Findley SD, Tamanaha M, Clegg NJ, Ruohola-Baker H. Maelstrom, a *Drosophila* spindle-class gene, encodes a protein that colocalizes with Vasa and RDE1/AGO1 homolog, Aubergine, in nuage. Development 2003;130:859–71.

[70] Lim AK, Kai T. Unique germ-line organelle, nuage, functions to repress selfish genetic elements in *Drosophila melanogaster*. Proc Natl Acad Sci USA 2007;104:6714–9.

[71] Aravin AA, Naumova NM, Tulin AV, Vagin VV, Rozovsky YM, Gvozdev VA. Double-stranded RNA-mediated silencing of genomic tandem repeats and transposable elements in the D. *melanogaster* germline. Curr Biol 2001;11:1017–27.

[72] Aravin AA, Klenov MS, Vagin VV, Bantignies F, Cavalli G, Gvozdev VA. Dissection of a natural RNA silencing process in the *Drosophila melanogaster* germ line. Mol Cell Biol 2004;24:6742–50.

[73] Nishida KM, Saito K, Mori T, Kawamura Y, Nagami-Okada T, Inagaki S, et al. Gene silencing mechanisms mediated by Aubergine piRNA complexes in *Drosophila* male gonad. RNA 2007;13:1911–22.

[74] Livak KJ. Detailed structure of the *Drosophila melanogaster* Stellate genes and their transcripts. Genetics 1990;124:303–16.

[75] Bozzetti MP, Massari S, Finelli P, Meggio F, Pinna LA, Boldyreff B, et al. The Ste locus, a component of the parasitic cry–Ste system of *Drosophila melanogaster*, encodes a protein that forms crystals in primary spermatocytes and mimics properties of the β-subunit of casein kinase 2. Proc Natl Acad Sci USA 1995;92:6067–71.

[76] Kotelnikov RN, Klenov MS, Rozovsky YM, Olenina LV, Kibanov MV, Gvozdev VA. Peculiarities of piRNA-mediated post-transcriptional silencing of Stellate repeats in testes of *Drosophila melanogaster*. Nucleic Acids Res 2009;37:3254–63.

[77] Nagao A, Mitsuyama T, Huang H, Chen D, Siomi MC, Siomi H. Biogenesis pathways of piRNAs loaded onto AGO3 in the *Drosophila* testis. RNA 2010;16:2503–15.

[78] Li C, Vagin VV, Lee S, Xu J, Ma S, Xi H, et al. Collapse of germline piRNAs in the absence of Argonaute3 reveals somatic piRNAs in flies. Cell 2009;137:509–21.

[79] Kawaoka S, Izumi N, Katsuma S, Tomari Y. 3′ end formation of PIWI-interacting RNAs in vitro. Mol Cell 2011;43:1015–22.

[80] Houwing S, Berezikov E, Ketting RF. Zili is required for germ cell differentiation and meiosis in zebrafish. EMBO J 2008;27:2702–11.

[81] Kawaoka S, Hayashi N, Suzuki Y, Abe H, Sugano S, Tomari Y, et al. The Bombyx ovary-derived cell line endogenously expresses PIWI/PIWI-interacting RNA complexes. RNA 2009;15:1258–64.

[82] Robine N, Lau NC, Balla S, Jin Z, Okamura K, Kuramochi-Miyagawa S, et al. A broadly conserved pathway generates 3′UTR-directed primary piRNAs. Curr Biol 2009;19:2066–76.

[83] Mi S, Cai T, Hu Y, Chen Y, Hodges E, Ni F, et al. Sorting of small RNAs into Arabidopsis argonaute complexes is directed by the 5′ terminal nucleotide. Cell 2008;133:116–27.

[84] Grimson A, Srivastava M, Fahey B, Woodcroft BJ, Chiang HR, King N, et al. Early origins and evolution of microRNAs and Piwi-interacting RNAs in animals. Nature 2008;455:1193–7.

[85] Aravin AA, Sachidanandam R, Girard A, Fejes-Toth K, Hannon GJ. Developmentally regulated piRNA clusters implicate MILI in transposon control. Science 2007;316:744–7.

[86] Eddy EM. Germ plasm and the differentiation of the germ cell line. Int Rev Cytol 1975;43:229–80.

[87] Vagin VV, Wohlschlegel J, Qu J, Jonsson Z, Huang X, Chuma S, et al. Proteomic analysis of murine Piwi proteins reveals a role for arginine methylation in specifying interaction with Tudor family members. Genes Dev 2009;23:1749–62.

[88] Nishida KM, Okada TN, Kawamura T, Mituyama T, Kawamura Y, Inagaki S, et al. Functional involvement of Tudor and dPRMT5 in the piRNA processing pathway in Drosophila germlines. EMBO J 2009;28:3820–31.

[89] Chen C, Jin J, James DA, Adams-Cioaba MA, Park JG, et al. Mouse Piwi interactome identifies binding mechanism of Tdrkh Tudor domain to arginine methylated Miwi. Proc Natl Acad Sci USA 2009;106:20336–41.

[90] Reuter M, Chuma S, Tanaka T, Franz T, Stark A, Pillai RS. Loss of the Mili-interacting Tudor domain-containing protein-1 activates transposons and alters the Mili-associated small RNA profile. Nat Struct Mol Biol 2009;16:639–46.

[91] Shoji M, Tanaka T, Hosokawa M, Reuter M, Stark A, Kato Y, et al. The TDRD9–MIWI2 complex is essential for piRNA-mediated retrotransposon silencing in the mouse male germline. Dev Cell 2009;17:775–87.

[92] Vasileva A, Tiedau D, Firooznia A, Müller-Reichert T, Jessberger R. Tdrd6 is required for spermiogenesis, chromatoid body architecture, and regulation of miRNA expression. Curr Biol 2009;19:630–9.

[93] Wang J, Saxe JP, Tanaka T, Chuma S, Lin H. Mili interacts with tudor domain-containing protein 1 in regulating spermatogenesis. Curr Biol 2009;19:640–4.

[94] Kirino Y, Kim N, de Planell-Saguer M, Khandros E, Chiorean S, Klein PS, et al. Arginine methylation of Piwi proteins catalysed by dPRMT5 is required for Ago3 and Aub stability. Nat Cell Biol 2009;11:652–8.

[95] Kirino Y, Vourekas A, Kim N, de Lima Alves F, Rappsilber J, Klein PS, et al. Arginine methylation of vasa protein is conserved across phyla. J Biol Chem 2010;285:8148–54.

[96] Patil VS, Kai T. Repression of retroelements in Drosophila germline via piRNA pathway by the tudor domain protein Tejas. Curr Biol 2010;20:724–30.

[97] Liu L, Qi H, Wang J, Lin H. PAPI, a novel TUDOR-domain protein, complexes with AGO3, ME31B and TRAL in the nuage to silence transposition. Development 2011;138:1863–73.

[98] Zamparini AL, Davis MY, Malone CD, Vieira E, Zavadil J, Sachidanandam R, et al. Vreteno, a gonad-specific protein, is essential for germline development and primary piRNA biogenesis in Drosophila. Development 2011;138:4039–50.

[99] Handler D, Olivieri D, Novatchkova M, Gruber FS, Meixner K, Mechtler K, et al. A systematic analysis of *Drosophila* TUDOR domain-containing proteins identifies Vreteno and the Tdrd12 family as essential primary piRNA pathway factors. EMBO J 2011;30:3977–93.

[100] Rangan P, Malone CD, Navarro C, Newbold SP, Hayes PS, Sachidanandam R, et al. piRNA production requires heterochromatin formation in *Drosophila*. Curr Biol 2011;21:1373–9.

[101] Olivieri D, Sykora MM, Sachidanandam R, Mechtler K, Brennecke J. An in vivo RNAi assay identifies major genetic and cellular requirements for primary piRNA biogenesis in *Drosophila*. EMBO J 2010;29:3301–17.

[102] Qi H, Watanabe T, Ku HY, Liu N, Zhong M, Lin H. The Yb body, a major site for Piwi-associated RNA biogenesis and a gateway for Piwi expression and transport to the nucleus in somatic cells. J Biol Chem 2011;286:3789–97.

[103] Zhang Z, Xu J, Koppetsch BS, Wang J, Tipping C, Ma S, et al. Heterotypic piRNA Ping-Pong requires qin, a protein with both E3 ligase and Tudor domains. Mol Cell 2011;44:572–84.

[104] Anand A, Kai T. The tudor domain protein kumo is required to assemble the nuage and to generate germline piRNAs in *Drosophila*. EMBO J 2011;31:870–82.

[105] Brennecke J, Malone CD, Aravin AA, Sachidanandam R, Stark A, Hannon GJ. An epigenetic role for maternally inherited piRNAs in transposon silencing. Science 2008;322:1387–92.

[106] Dalmay T, Horsefield R, Braunstein TH, Baulcombe DC. SDE3 encodes an RNA helicase required for post-transcriptional gene silencing in Arabidopsis. EMBO J 2001;20:2069–77.

[107] Meister G, Landthaler M, Peters L, Chen PY, Urlaub H, Lührmann R, et al. Identification of novel argonaute-associated proteins. Curr Biol 2005;15:2149–55.

[108] Szakmary A, Reedy M, Qi H, Lin H. The Yb protein defines a novel organelle and regulates male germline stem cell self-renewal in *Drosophila melanogaster*. J Cell Biol 2009;185:613–27.

[109] Choi SY, Huang P, Jenkins GM, Chan DC, Schiller J, Frohman MA. A common lipid links Mfn-mediated mitochondrial fusion and SNARE-regulated exocytosis. Nat Cell Biol 2006;8:1255–62.

[110] Watanabe T, Chuma S, Yamamoto Y, Kuramochi-Miyagawa S, Totoki Y, Toyoda A, et al. MITOPLD is a mitochondrial protein essential for nuage formation and piRNA biogenesis in the mouse germline. Dev Cell 2011;20:364–75.

[111] Pohlman RF, Liu F, Wang L, Moré MI, Winans SC. Genetic and biochemical analysis of an endonuclease encoded by the IncN plasmid pKM101. Nucleic Acids Res 1993;21:4867–72.

[112] Nishimasu H, Ishizu H, Saito K, Fukuhara S, Kamatani MK, Matsumoto N, et al. Structure and function of Zucchini endoribonuclease in piRNA biogenesis. Nature 2012; in press.

[113] Ipsaro JJ, Haase AD, Simon R, Knott SR, Joshua-Tor L, Hannon GJ. The structural biochemistry of Zucchini implicates it as a nuclease in piRNA biogenesis. Nature 2012; in press.

[114] Stuckey JA, Dixon JE. Crystal structure of a phospholipase D family member. Nat Struct Mol Biol 1999;6:278–84.

[115] Gottlin EB, Rudolph AE, Zhao Y, Matthews HR, Dixon JE. Catalytic mechanism of the phospholipase D superfamily proceeds via a covalent phosphohistidine intermediate. Proc Natl Acad Sci USA 1998;95:9202–7.

[116] Nagao A, Sato K, Nishida KM, Siomi H, Siomi MC. Gender-specific hierarchy in nuage localization of PIWI-interacting RNA factors in *Drosophila*. Front Genet 2011;2:55.

[117] Klattenhoff C, Xi H, Li C, Lee S, Xu J, Khurana JS, et al. The *Drosophila* HP1 homolog Rhino is required for transposon silencing and piRNA production by dual-strand clusters. Cell 2009;138:1137–49.

[118] Chen Y, Pane A, Schüpbach T. *cutoff* and *aubergine* mutations result in retrotransposon upregulation and checkpoint activation in *Drosophila*. Curr Biol 2007;17:637–42.

[119] Pane A, Jiang P, Zhao DY, Singh M, Schüpbach T. The Cutoff protein regulates piRNA cluster expression and piRNA production in the *Drosophila* germline. EMBO J 2011;30:4601–15.

[120] Sato K, Nishida KM, Shibuya A, Siomi MC, Siomi H. Maelstrom coordinates microtubule organization during *Drosophila* oogenesis through interaction with components of the MTOC. Genes Dev 2011;25:2361–73.

[121] Specchia V, Piacentini L, Tritto P, Fanti L, D'Alessandro R, Palumbo G, et al. Hsp90 prevents phenotypic variation by suppressing the mutagenic activity of transposons. Nature 2010;463:662–5.

[122] Gangaraju VK, Yin H, Weiner MM, Wang J, Huang XA, Lin H. *Drosophila* Piwi functions in Hsp90-mediated suppression of phenotypic variation. Nat Genet 2011;43:153–8.

[123] Preall JB, Czech B, Guzzardo PM, Muerdter F, Hannon GJ. *shutdown* is a component of the *Drosophila* piRNA biogenesis machinery. RNA 2012;18:1446–57.

[124] Barbosa V, Kimm N, Lehmann R. A maternal screen for genes regulating Drosophila oocyte polarity uncovers new steps in meiotic progression. Genetics 2007;176:1967–77.

CHAPTER EIGHT

Control of MicroRNA Maturation by p53 Tumor Suppressor and MCPIP1 Ribonuclease

Hiroshi I. Suzuki, Kohei Miyazono[1]

Department of Molecular Pathology, Graduate School of Medicine, The University of Tokyo, Tokyo, Japan
[1]Corresponding author: e-mail address: miyazono@m.u-tokyo.ac.jp

Contents

Abstract

MicroRNAs (miRNAs) are small noncoding RNAs that function as major players of post-transcriptional gene regulation in diverse species. In mammals, the biogenesis of miRNAs is executed by cooperation of multiple biochemical reactions including processing of miRNA precursors by two central endoribonucleases, Drosha and Dicer. While the transcription of miRNA precursors is regulated by various transcription factors, it has been recently unlabeled that miRNA maturation is further subjected to a range of regulatory mechanisms in miRNA processome. We previously demonstrated that a key regulator of tumor suppression, p53, modulates Drosha-mediated processing of primary miRNA transcripts. The pathways for miRNA maturation also dynamically cross talk with other intracellular signaling pathways. Further, we recently revealed that MCPIP1, an immune regulator, antagonizes Dicer and suppresses miRNA processing as another

The Enzymes, Volume 32
ISSN 1874-6047
http://dx.doi.org/10.1016/B978-0-12-404741-9.00008-8
163

important endoribonuclease. In addition, recent evidences highlighted alterations of miRNA biogenesis in cancer pathogenesis. These advances in miRNA metabolome would bring deep insights for understanding alterations of miRNAs in diverse disease pathogenesis and for development of miRNA-based therapeutic applications. In this chapter, we summarize our findings and advances in miRNA metabology.

1. INTRODUCTION

MicroRNAs (miRNAs) are single-strand RNAs with 21–25 nucleo-tides (nt) length and abundant class of small RNA species. miRNAs function as posttranscriptional gene regulators in cooperation with RNA-induced silencing complex (RISC) and modulate a range of mRNAs to generally suppress their protein output [1]. Recognition of target mRNAs by miRNAs is mediated through complementary base pairing between indi-vidual miRNAs and their target mRNAs, and the 5′ seed region of miRNAs (2–7 nucleotides) is especially important for target selection in animals [2]. Reflecting the dependence on this short sequence for target recognition, mammalian individual miRNAs modulate the destiny of hundreds of mRNAs, thereby enabling widespread commitment in diverse cellular func-tions including differentiation, motility, proliferation, and cell death.

miRNAs are involved in various disease pathogenesis and are extensively researched especially in cancer research field [3]. While miRNAs can function as both tumor suppressors and tumor promoters in tumor progression, early reports revealed that overall miRNA levels are frequently downregulated in tumor tissues and implicated their intrinsic tumor-suppressive functions [4,5]. Kumar *et al.* demonstrated that Dicer, a central ribonuclease in miRNA biogenesis, functions as a haploinsufficient tumor suppressor in mouse cancer models [6,7]. Other report also showed that monoallelic but not biallelic loss of Dicer promotes tumorigenesis *in vivo* in a retinoblastoma-sensitized background [8], suggesting that miRNA pathway displays characteristics of haploinsufficient tumor suppression system. In addition, accumulating evidences have indicated alterations of miRNA biogenesis at multiple steps of miRNA maturation in cancer [4,9].

Another line of evidences have disclosed the complex regulatory landscape governing intracellular miRNA biosynthesis [10,11]. While the transcription of diverse miRNA precursors is regulated by various transcription factors, it has been shown that miRNA maturation is further subjected to a range of regulatory mechanisms. We previously demonstrated that a key regulator of tumor suppression, p53, modulates Drosha-mediated pri-miRNA processing [12]. In accordance with our findings, miRNA biogenesis pathways dynamically cross talk with other intracellular signaling pathways. Further, many reports showed the involvement of various RNA-binding proteins in miRNA species-specific regulation of miRNA maturation. In addition, we recently revealed that MCPIP1, an immune regulator, antagonizes Dicer and suppresses miRNA processing as another important endoribonuclease [13]. This finding suggests that the balance between processing and destructing ribonucleases modulates miRNA biosynthesis. In this chapter, we summarize our findings in miRNA metabology and accumulating evidences in miRNA processing. We mainly focus on the findings in mammalian system.

2. TWO-STEP ENDORIBONUCLEASE REACTIONS IN miRNA MATURATION

2.1. Scheme of miRNA processing

Two RNase III enzymes, Drosha and Dicer, have essential roles in miRNA maturation in mammalian cells [10,11]. While RNA polymerase III is involved in the transcription of primary miRNA transcripts (pri-miRNAs) of some miRNA genes, majority of miRNA genes are first transcribed to pri-miRNAs by RNA polymerase II (Fig. 8.1, step 1). As well as other protein-coding mRNAs, the expression levels of these pri-miRNAs are regulated by various transcription factors including tumor suppressive and oncogenic transcription factors. In the next step (Fig. 8.1, step 2), the pri-miRNAs are catalyzed in the nucleus by Drosha and DGCR8 (DiGeorge critical region 8), a double-stranded RNA-binding domain (dsRBD) protein, which comprise the Microprocessor complex. Drosha cleaves the pri-miRNAs at the site of pri-miRNAs one helical (~11 bp) turn away from the single-stranded RNA (ssRNA)–double-stranded RNA (dsRNA) junction, producing hairpin-structured precursor miRNAs (pre-miRNAs) of about 60–70 nt. DGCR8 recognizes ssRNA–dsRNA junction and regulates Drosha cleavage sites [14]. To make biologically active mature miRNAs, the place of miRNA maturation transits from the nucleus to the cytoplasm (Fig. 8.1, step 3). Exportin-5 (XPO5) binds pre-miRNAs and shuttles them into the cytoplasm in concert with GTP-bound form of the Ran cofactor [15]. The pre-miRNAs are further processed into double-stranded miRNA duplex of 20–22 nt by cytosolic endoribonuclease Dicer (Fig. 8.1, step 4). Other dsRBD protein TRBP/PACT supports Dicer

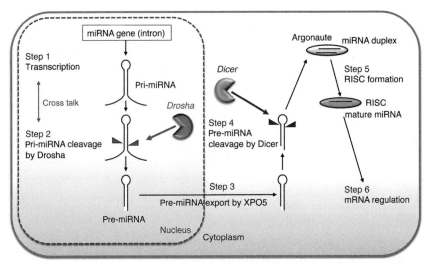

Figure 8.1 Scheme of miRNA biogenesis. In mammalian cells, miRNA maturation occurs through multiple biochemical steps, including pri-miRNA transcription (Step 1), Drosha-mediated cleavage of pri-miRNA to pre-miRNA (Step 2), pre-miRNA export by XPO5 (Step 3), Dicer-mediated cleavage of pre-miRNA to miRNA duplex (Step 4), and RISC formation (Step 5). Resultant RISC controls the protein output of target mRNAs (Step 6). (For color version of this figure, the reader is referred to the online version of this chapter.)

processing. One strand of miRNA duplex is then stably incorporated as a mature miRNA into a member of the Argonaute family of proteins, the core of RISC, together with several cofactors (Fig. 8.1, step 5). In mammals, four Argonaute proteins (Ago1, Ago2, Ago3, and Ago4) have been identified. At a final step (Fig. 8.1, step 6), the resulting RISC mainly suppresses the protein output of numerous target mRNAs based on the selection of target mRNAs through the sequence complementarity between miRNAs and target mRNAs. While this maturation process can be applied to majority of miRNAs, alternative processing pathways for some miRNAs including the mirtron pathway [16,17] and the Ago2-mediated miR-451 processing pathway [18,19] deviate from the conventional miRNA biogenesis pathway and do not necessitate Drosha or Dicer.

2.2. Cofactors of Drosha complex

The Microprocessor complex composed of Drosha and DGCR8 forms a large protein complex in association with multiple auxiliary factors [20]. Various factors have previously been identified as cofactors of protein complexes containing overexpressed Drosha: DEAD-box RNA helicases (DDX1, DDX3X, p68 (DDX5), and p72 (DDX17)), DEAH-box RNA helicase (DHX15), dsRNA-binding proteins (NF90 (nuclear factor 90) and NF45), heterogeneous nuclear ribonucleoproteins (hnRNPs) (hnRNPU-like, hnRNPUb, hnRNPM4, hnRNPH1, hnRNPD-L, hnRNP RALY, and TDP-43), the Ewing's sarcoma family of proteins containing an RNA recognition motif and a zinc-finger domain (EWS, FUS/TLS, and TAF15), and protein kinase SPRK1a [20]. In another report, 11 DGCR8-associated proteins have also been identified: nucleolin, heat-shock proteins (Bip and Hsp70), RNA helicases (p68, p72, and DHX9), hnRNPs (hnRNP R, hnRNP H1, and hnRNPU), FUS/TLS, and ILF3 [21].

Many of these proteins contain various RNA-binding domains and play roles in RNA metabolism, and the roles in miRNA maturation have been investigated for some members. DEAD-box RNA helicases modulate RNA structures through ATP binding and hydrolysis. p68 and p72 DEAD-box RNA helicases mediate several RNA regulatory functions, including transcription, splicing, RNA degradation, RNA export, and translation [22,23]. In the context of miRNA maturation, p68 and p72 positively control the recruitment of the Drosha complex to some pri-miRNAs and the subsequent pri-miRNA cleavage, thereby allowing the efficient maturation of a subset of miRNAs such as miR-15, -16, and -145 [24]. While NF90 and NF45 do not seem to associate with intrinsic Drosha, NF90/NF45 complex interacts with some pri-miRNAs and suppresses Drosha-mediated pri-miRNA cleavage [25]. Among hnRNPs, hnRNP L and other hnRNP, hnRNP A1, modulate pri-miRNA processing of several miRNAs through specific interaction with a subset of pri-miRNAs [26]. Two genes of Drosha cofactors, TDP-43 and FUS/TLS, are known to be mutated in amyotrophic lateral sclerosis and frontotemporal lobar degeneration, and it has recently been shown that TDP-43 promotes miRNA biogenesis for some miRNAs as cofactors of Drosha and Dicer complexes [27]. Further, several factors including KSRP, Mll/Af4, Mll/Af9, SNIP1, SF2/ASF, and Ars2 have been reported as other Drosha cofactors and modulators of pri-miRNA processing [28–33].

The link between these RNA-associated proteins and diverse RNA regulatory functions, summarized as follows, also suggests important implication in pri-miRNA processing: (1) p68/p72: regulation of transcription, splicing, RNA degradation, and RNA export; (2) NF90/NF45: transcription regulation; (3) hnRNPs: splicing regulation; (4) TDP-43: splicing regulation; (5) KSRP: regulation of splicing and RNA decay; (6) Mll/Af4

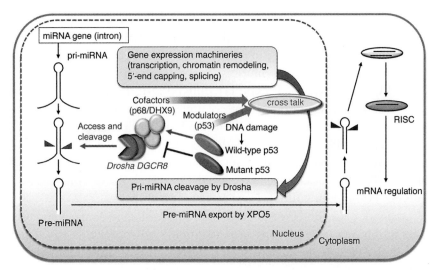

Figure 8.2 Regulation of miRNA processing by p53. Cross talk between pri-miRNA processing and nuclear gene expression machineries have been highlighted by recent studies. In this context, p53 modulates Drosha-mediated miRNA processing. (For color version of this figure, the reader is referred to the online version of this chapter.)

and Mll/Af9: transcription regulation; (7) SNIP1: transcription regulation; (8) SF2/ASF: splicing regulation; and (9) Ars2: control of 5′-end capping. This namely implicates dynamic coupling between miRNA processing and several nuclear functional instruments executing transcription, 5′-end capping, and splicing (Fig. 8.2) [34].

It has been shown that key events generating pri-miRNAs and mRNAs, such as transcription, splicing, and processing are spatiotemporally coupled in the nucleus [34]. Drosha cleaves intronic pri-miRNAs between the splicing commitment step and the excision step, and not after the completion of splicing catalysis [35]. The functional association of Drosha with splicing factors and spliceosome supports these findings [31,36]. Other reports have demonstrated that Drosha is recruited cotranscriptionally and cleaves pri-miRNAs during transcription [37,38]. Importantly, Pawlicki *et al.* showed that increased retention of the nascent pri-miRNAs at the site of transcription through deletion of 3′-end processing signals or use of a miRNA promoter leads to three- to fourfold greater efficiency in pri-miRNA processing [39]. In addition, the intrinsic pri-miRNAs that undergo efficient processing are enriched in chromatin-associated nuclear fractions [37,39]. These findings indicate that pri-miRNAs retained at transcription sites are processed more efficiently than released pri-miRNAs and strongly suggest that recruitment of Drosha to nascent pri-miRNAs at the transcription site may promote entry into pri-miRNA processing. Thus, many Drosha cofactors involved in other RNA processing pathways may modulate pri-miRNA processing by targeting this important link between pri-miRNA production (transcription, splicing, and so on) and Drosha cleavage, leading to control of the fidelity and activity of Drosha. It suggests that alterations in transcription factor activities, chromatin structures, and splicing activities may directly influence Drosha processing. This may be also emphasized by the fact that several proteins, including EWS, FUS/TLS, TAF15, and SRPK1a are involved in transcription or cotranscriptional processes [34], while the roles of these proteins in miRNA maturation have not yet been reported.

3. CONTROL OF miRNA BIOSYNTHESIS BY p53

3.1. p53 regulates miRNA processing through interaction with Drosha complex

Cross talk between pri-miRNA processing and nuclear gene expression machineries have been highlighted by several reports of posttranscriptional regulation of Drosha cleavage by transcriptional factors [4,11]. We previously demonstrated that a key tumor suppressor, p53, controls miRNA maturation of a subset of miRNAs (Fig. 8.2) [12], opening the avenue for understanding complexity of miRNA metabolome.

In response to various environmental changes, the expression profiles of cellular miRNAs display dynamic changes. DNA damage encompasses dynamic gene expression at the transcriptional and posttranslational levels, provoking DNA repair, cell-cycle arrest, and apoptosis. We observed that certain miRNAs such as miR-143 and miR-16 increased under DNA damage induction by the treatment with the DNA-damaging agent doxorubicin [12]. Kinetic analyses revealed that these miRNAs, including miR-16-1, miR-143, miR-145, and miR-206, showed no significant change in pri-miRNA expression levels but showed upregulation of pre-miRNAs and mature miRNAs, suggesting that a subset of miRNAs is induced by posttranscriptional mechanisms probably at the Drosha cleavage step. In contrast, DNA damage induced the expression of miR-34a, a transcriptional target of p53, at all layers of pri-miRNA, pre-miRNA, and mature miRNA levels. Consistently, Pothof *et al.* also reported posttranscriptional expression change of miRNAs, including miR-16 induction, during UV-induced DNA damage response [40]. In the context of transcriptional regulation, p68 interacts with components of the transcription factors such as RNA polymerase II and CBP/p300 and works as a transcriptional coregulator for transcription factors, including p53 and estrogen receptor α (ERα) [22,41,42]. On the basis of these findings, we investigated potential involvement of p53, p68, and p72 and found that p53 and p68/p72 are necessary for this upregulation of mature miRNAs under DNA damage, indicating dual roles of p53 in miRNA expression control, that is, transcription-dependent miR-34a upregulation and enhancement of miRNA maturation. It was also supported by comparison of miRNA profiling data showing that p72-dependent miRNAs tended to increase during DNA damage induction. In the report by Pothof *et al.*, this posttranscriptional upregulation of miRNAs was attenuated in the cells compromised for p53 function [40].

In accordance with the previous report of interaction between p53 and p68/p72 [41], we subsequently observed that endogenous p53 accumulated and interacted with the Drosha/DGCR8 complex, depending on the presence of RNAs and the p68/p72 RNA helicases, upon DNA damage induction in human cells. In an *in vitro* cleavage analysis, association between Drosha complex and p53 promoted the activities of Drosha to cleave pri-miRNAs. Interestingly, RNA immunoprecipitation analyses revealed that p53 augmented the recruitment of Drosha and p68 to pri-miRNAs in a DNA damage-dependent manner. Considering the coupling of miRNA processing with transcriptional machineries, p53 might form a ternary complex with Drosha/DGCR8 and p68/p72 and bridge the nascent pri-miRNAs and miRNA processing factory, leading to facilitation of Drosha-mediated pri-miRNA processing. In addition, these miRNAs, which are posttranscriptionally upregulated by p53, suppressed *in vitro* cell proliferation and targeted important regulators of the cell cycle and cell proliferation including K-Ras and cyclin-dependent kinase (CDK) 6, suggesting that the enhanced miRNA processing might complement p53 activities [12].

In human cancer, p53 is one of the most frequently altered tumor suppressor genes. Mutation in cancer-derived p53 has been shown to drive additional oncogenic activities such as the facilitation of metastasis, suppression of other p53 family members, and inhibition of ATM signaling [43–46]. We observed that these transcriptionally inactive cancer-derived p53 mutants suppressed biosynthesis of the above-mentioned miRNAs. Introduction of mutant p53 attenuated the association between Drosha/DGCR8 and p68/p72 and compromised the recruitment of Drosha and p68 to pri-miRNAs. Thus, in the context of cotranscriptional miRNA processing, mutant p53 may interfere with efficient association between pri-miRNAs and Drosha and their cofactors in contrast to wild-type p53. Our findings demonstrated another important aspect of this key tumor suppressor [12].

3.2. Modulation of pri-miRNA processing by other transcriptional regulators

In parallel studies, it has been demonstrated that the Drosha complex is influenced by various transcriptional regulators. In the signaling of trans-forming growth factor-β or bone morphogenetic proteins, the key signal transducers Smads enhanced Drosha-mediated cleavage of pri–miR-21 into pre-miR-21 [47]. In addition, a stem cell factor, Nanog, interacts with Drosha–p68 complex, promotes the production of miR-21, and confers chemotherapy resistance in breast cancer cells [48]. On the other hand, ERα, a partner of p68/p72, has been shown to inhibit maturation of several miRNAs, which overlapped with the target miRNAs of p53-mediated bio-synthesis modulation [49].

Most recently, Kawai *et al.* unveiled that the tumor suppressor breast can-cer 1 (BRCA1) accelerates the processing of pri–miRNAs through associ-ation with Drosha complex [50]. BRCA1 forms a large protein complex known as the BRCA1-associated genome surveillance complex and also functions as a transcriptional modulator through regulation of chromatin structure together with RNA polymerase II and histone deacetylase com-plexes. In their report, BRCA1 interacts with Drosha/DGCR8, p68, DHX9, another cofactor of Drosha, and p53/Smad, respectively, and fur-ther directly binds to pri-miRNAs [50]. This report showed that BRCA1 facilitates Drosha function in concert with DHX9 DEAH-box RNA heli-case. These findings have demonstrated that several nuclear factors modulate miRNA biogenesis in addition to their roles in transcriptional regulation. Therefore, p68/p72 and other RNA-binding molecules might function as an interface integrating intracellular signaling pathways into modification of miRNA processome (Fig. 8.2).

3.3. Cross talk between p53 network and miRNA system

Previous studies continuously showed a close relationship between p53 pathway and miRNA network [4]. As described above, p53 modulates the expression of various miRNAs through several mechanisms. As a tran-scriptional activator, p53 activates the transcription of miR-34 family and induces the expression levels of these miRNAs [51]. miR-34s target multiple genes including CDK4, cyclin E2, and Bcl-2 and control cell cycle and apoptosis. In addition, p53 upregulates certain miRNAs such as miR-192, miR-194, and miR-215 [52,53]. On the other hand, p53 suppresses miR-17-92 oncogenic miRNA cluster through transcriptional repression. In hypoxic condition, the expression of miR-17-92 cluster is suppressed by a competitive inhibition of the recruitment of TATA-binding protein to the promoter of miR-17-92 gene by p53 [54]. Another line of evidences also placed miRNAs upstream regulators and modulators of the p53 pathways [4]. It has been reported that multiple miRNAs such as miR-29 family, miR-125b, miR-122, and miR-372/373 control activities of p53 and the p53 downstream pathways [55–58].

Our study suggests widespread interaction between p53 network and the miRNA system, as overall miRNA system has been postulated to play intrinsic tumor-suppressive function. This notion may be supported by the finding that a loss of mature miRNAs results in increased DNA damage and p53 activity [59], revealing a reciprocal connection between the p53 and miRNA pathways. Further, Su *et al.* have identified Dicer and miR-130b as transcriptional targets of p63, a p53 family member, but not p53, underscoring the cross talk between p53 family members and miRNAome [60]. Thus, the p53 tumor-suppressive pathways may have intrinsic property to maintain cellular miRNA expression levels for tumor suppression by targeting multiple points of the miRNA biosynthesis.

4. NEGATIVE REGULATION OF miRNA PROCESSING BY MCPIP1

4.1. Turnover of miRNA precursors and miRNAs

Research advances in the past 5 years provide multiple insights into the regulation of miRNA production, which are represented by (1) extensive cross talk between miRNA maturation pathways and intracellular signaling pathways and (2) biased processing of diverse miRNA species through pairing between individual miRNAs and specific RNA-binding proteins. For example, latter mechanisms include let-7 regulation by Lin28. In contrast, the mechanisms of modification, turnover, stabilization, and reduction of miRNAs are still largely unknown [61].

In general, miRNAs are thought to be stable molecules. In the study by Gantier *et al.*, the half-lives of miRNAs were ranged from 28 to 200 h and were longer than those of mRNAs [62]. It could be explained by

Argonaute-mediated protection from promiscuous ribonucleases, as Argonautes have been shown to enhance the overall abundance of miRNAs [63]. On the other hand, it has been also reported that individual miRNAs are subjected to rapid degradation in a miRNA species-specific manner and in certain specific cellular contexts, such as cell-cycle transition and dark adaptation in neurons [64–66].

Several nucleases, such as SDN1 (small RNA degrading nuclease 1) in *Arabidopsis* and XRN2 and XRN1 in *Caenorhabditis elegans* have been so far reported to destruct miRNAs [67–69]. SDN1 and XRN2/XRN1 degrade mature miRNAs in a $3' \rightarrow 5'$ direction and $5' \rightarrow 3'$ direction, respectively. While the details of miRNA degradation mechanisms have not been well investigated in mammalian systems, the involvement of several exoribonucleases including XRN1, PNPT1, and RRP41 has been proposed in the studies using mammalian cells [70,71]. In contrast to XRN1, PNPT1 and RRP41 are components of exosome, which is $3' \rightarrow 5'$ exoribonuclease complex. However, it remains largely unknown which mechanisms contribute to the specificity of target miRNAs and how the degradation processes by these nucleases is actively regulated in the context of specific cellular conditions. The mechanisms for turnover of pri-miRNAs and pre-miRNAs also remain to be elucidated.

4.2. Negative regulation of pre-miRNA processing by MCPIP1 ribonuclease

It is not well understood whether the processing by Drosha and Dicer are rate-limiting steps actively inhibited by endogenous suppressor systems. We recently identified MCPIP1 as a representative molecule responsible for such notion and showed that not only mature miRNAs but also pre-miRNAs are targets of active degradation to reduce miRNA activities (Fig. 8.3) [13].

While miRNA and RNA interference (RNAi) pathways are well conserved across a range of species, previous studies showed certain evolutional differences, in particular, from immunological aspects, as represented by different core antiviral mechanisms between plants/invertebrates and mammals, that is, respectively, RNAi response and interferon (IFN) response [72]. In mammalian system, previous studies have established the close relationship between the immune system and the small RNA regulation. Expression of miRNAs dynamically changed during the immune reaction [73,74]. As for regulated miRNA biosynthesis, KSRP has been reported to regulate maturation of miR-155 during macrophage activation [75]. In addition,

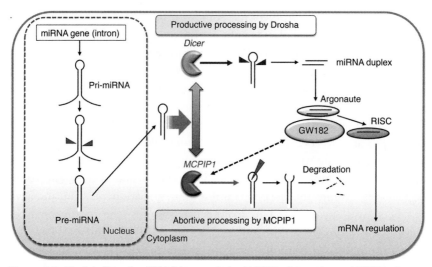

Figure 8.3 Modulation of miRNA biogenesis by MCPIP1 ribonuclease. MCPIP1 ribonu-clease cleaves the terminal loops of pre-miRNA, antagonizes Dicer processing, thereby limiting miRNA biogenesis. (For color version of this figure, the reader is referred to the online version of this chapter.)

PNPT1 (also known as hPNPase^{old-35}) is one of type I IFN-inducible genes, and Swadesh *et al.* reported that PNPT1 mediates IFN-induced miR-221 downregulation by degrading miR-221 [71], indicating that mammalian immune system really affects miRNA metabolome.

By taking these observations into account, we screened novel regulators of the miRNA pathway with particular focus on the immune system–associated genes containing a potential RNA-binding domain and identified MCPIP1 (also known as Zc3h12a), one of CCCH-type zinc-finger proteins with RNA-binding potential [76,77], as a potent suppressor of miRNA activity and production. Further analyses demonstrated that overexpressed MCPIP1 localized in the cytoplasm in association with GW182, destructs pre-miRNAs, and antagonizes Dicer processing, leading to inhibition of *de novo* miRNA maturation. Silencing of MCPIP1 caused upregulation of overall miRNA expression levels, suggesting that MCPIP1 functions as a broad suppressor of miRNA biogenesis. The N-terminus of MCPIP1, termed Nedd4-BP1, bacterial YacP Nuclease (NYN) domain, is well conserved among other related proteins including MCPIP2, MCPIP3, and MCPIP4 and has been predicted to show nuclease function [78]. In consistent with a previous report showing that MCPIP1/Zc3h12a mediates the decay of IL-6 mRNA [79], this domain was essential for nuclease activity for pre-miRNAs. We further found that MCPIP1 contains several additional functional domains important for inhibition of miRNA maturation: (1) CCCH zinc-finger motif mediates interaction with pre-miRNAs and (2) vertebrate-specific C-terminus controls oligomerization of MCPIP1 and interaction with pre-miRNAs. While the details of recognition of target pre-miRNAs by MCPIP1 remains to be elucidated, *in vitro* cleavage analyses demonstrated that MCPIP1 preferentially cleaves the unpaired regions around the terminal loops of pre-miRNAs as an endoribonuclease. Pre-miRNAs and target region of IL-6 mRNA form similar hairpin structures, suggesting that MCPIP1 modulates diverse RNA species containing hairpin structures as Drosha cleaves both pri-miRNAs and the hairpin structure within DGCR8 mRNA [80].

In addition to PNPT1, MCPIP1 might be a potential player linking immune reaction and miRNA metabolism [81]. MCPIP1 is dynamically induced by inflammatory stimuli, including lipopolysaccharide (LPS) and IL-1β, in a Myd88-dependent fashion [76,77,79]. Two groups reported that $MCPIP1^{-/-}$ mice showed growth retardation, lymphadenopathy, splenomegaly, and enhanced inflammatory symptoms [79,82], emphasizing its biological importance. Besides the roles of ribonuclease targeting several cytokine mRNAs, Liang *et al.* reported that MCPIP1 deubiquitinates TRAF proteins and negatively regulates JNK and NF-κB-signaling pathways [82]. Severe phenotype in $MCPIP1^{-/-}$ mice suggests additional targets and biological roles of MCPIP1, and we consistently observed that MCPIP1 suppresses miR-155/c-Maf/IL-4 axis, limits miR-155 upregulation during innate immune response, and mediates miR-16 downregulation upon macrophage differentiation [13], suggesting that MCPIP1 may be involved in altered miRNA metabolism during immune response. Collectively, our findings demonstrate that the balance of productive and abortive ribonucleases modulates miRNA biogenesis (Fig. 8.3). Our study also implies the involvement of not only exonucleases but also additional endoribonucleases in the miRNA maturation pathways.

Our findings might be also discussed from the evolutional standpoint. Poor conservation of MCPIP1 C-terminus suggests that vertebrate MCPIP family has diversified and evolved anti-miRNA function of MCPIP1, which is functionally similar to suppressor of RNA silencing (SRS) developed by plant/invertebrate viruses against antiviral RNAi response [72]. While plants and invertebrates operate viral-derived siRNA and RNAi pathway as cardinal components of antiviral defense, the conventional RNA silencing machinery in mammals appears to be more specialized for posttranscriptional gene regulation and transfer the major role in antiviral defense to the so-called IFN system [72]. On the other hand, miRNAs encoded in certain mammalian viruses perturb host gene regulatory networks [72,83] and there exist several mechanisms represented by RNase L and ZAP, another member of CCCH-type zinc-finger proteins, to avoid the promiscuity by viral-derived RNA species in vertebrate cells [84,85]. On reflection, these evolutional differences of RNA silencing and emergence of endogenous SRS-like molecule, MCPIP1, further envisage that the vertebrate miRNA system might transit into an unassailable rather than flexible architecture against small RNA promiscuity, as potentially caused by viral-derived miRNAs, along the evolution of IFN system and its downstream destruction strategy of foreign RNA. It may be contrasting to the biologically flexible use of viral-derived small RNAs in antiviral RNAi response in plants and invertebrates and counterrevolution of SRSs by viruses. MCPIP1 might thus function to preserve the integrity of endogenous miRNA system by limiting *de novo* miRNA biogenesis in vertebrate cells.

5. AUTOREGULATION OF miRNA PROCESSING PROCESSES

Although transcriptional regulation of pri-miRNAs is important for regulation of miRNA expression levels, array of evidences summarized in this chapter clearly suggest that miRNA processing is a very complex and delicate process that multiple regulatory machineries can converge on. To maintain homeostasis of intracellular miRNA generation, autoregulatory feedback system is implemented within this miRNA biosynthesis machineries. One is an auto-feedback loop between Drosha and DGCR8 [80], in which DGCR8 stabilizes Drosha protein level, while Drosha destabilizes DGCR8 mRNA through cleavage of the hairpin structures in the DGCR8 mRNA. Omer *et al.* recently showed that DGCR8 mRNA is an inferior Drosha substrate than pri-miRNAs and that increase in overall pri-miRNA expression enhanced the levels of the Microprocessor complex by antagonizing DGCR8 mRNA cleavage by Drosha [86], suggesting that mammalian cells can finely adjust the levels of Microprocessor according to pri-miRNA levels to maintain the efficiency and specificity of Microprocessor for pri-miRNAs over other nuclear RNAs. In addition, Bennasser *et al.* reported cross-regulation between XPO5, pre-miRNAs, and Dicer. They found that XPO5 transports Dicer mRNA to maintain Dicer protein levels and that competition for XPO5 binding between pre-miRNAs, viral small

RNAs, and Dicer mRNA negatively regulates Dicer levels [87]. It is contrasting to the case of Drosha, in which the processing capacity of pri-miRNAs is robust to pri-miRNA levels, and it might be reasoned that this type of regulation of Dicer might be necessary to prevent adverse reactions provoked by accumulation of excess amounts of small RNA species in the cytoplasm, thereby leading to the different strategy in the nucleus and in the cytoplasm. It was consistent with the consideration described in the previous section, in which mammalian system may contain several limiter systems for miRNA biogenesis against small RNA promiscuity.

6. ALTERATION OF miRNA BIOGENESIS IN HUMAN CANCER

miRNAs show aberrant expression pattern in diverse disease pathogenesis, and recent advances reveal frequent alteration of miRNA biogenesis itself in human cancer. Several mechanisms including downregulation of miRNA processing factors such as Dicer and Drosha, and mutations of TRBP and XPO5 have been reported (Fig. 8.4) [88–92]. Dicer1 mutations are observed in familial pleuropulmonary blastoma and in nonepithelial ovarian

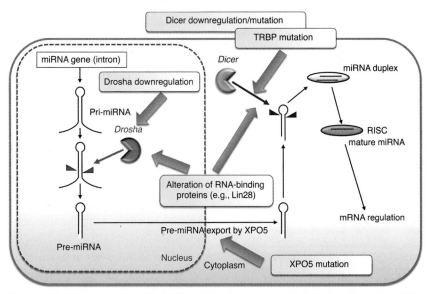

Figure 8.4 Alteration of miRNA biogenesis in cancer. Several alterations in miRNA processing including downregulation of Dicer and Drosha, and mutations of TRBP and XPO5 have been reported in human cancer. (For color version of this figure, the reader is referred to the online version of this chapter.)

cancers as germline mutation and somatic mutation [93,94], respectively. Accordingly, it was reported that reduced expression of Dicer is associated with poor prognosis in lung cancer and ovarian cancer [88,89]. In the previous report, we examined the relationship between Dicer, MCPIP1, and cancer prognosis using the transcriptome data of lung cancer patients and observed that MCPIP1 and Dicer are in antagonistic relationship and that high expression of MCPIP1 is associated with poor clinical prognosis in an opposite manner to Dicer [13]. While further investigation is important, these findings raised the scenario that MCPIP1 upregulation by several inflammatory stimuli might be partly attributable to alteration of miRNA biogenesis in cancer, in addition to reduced expression of Dicer and/or Drosha and mutations in TRBP and XPO5.

7. CONCLUSION

In this chapter, we summarize our findings on regulation of miRNA maturation by p53 and MCPIP1 and current advances in miRNA metabology. These discoveries have provided conceptual advances in the regulation of small RNA world. These findings may be important for not only understanding of pathological dysregulation of miRNA expression and function but also development of miRNA-based therapeutic interventions. Recent studies recurrently reported alterations of miRNA biogenesis in human cancer [4,9]. One may postulate that the miRNA maturation processes themselves may not be good candidates for therapeutic targeting, because such strategies modulating miRNA biosynthesis may affect diverse miRNA species, including both tumor-suppressive miRNAs and oncogenic miRNAs. However, a most recent study has presented the possibility that modulation of not only individual miRNAs but also miRNA processing machinery could aid in cancer treatment. Melo *et al.* recently showed that small molecule enoxacin, a chemical enhancer of RNAi, exerts cancer-specific growth inhibition by enhancing TRBP-mediated miRNA processing [95]. In addition, this research field would enhance our understanding about the responses of living system to nucleic acid medicine and provide insights for improvement of rational therapeutic design. Further innovation in miRNA metabolome and RNA interactome will offer a biological basis for diagnostic and therapeutic strategies based on RNA biology.

ACKNOWLEDGMENTS

We thank members of Department of Molecular Pathology, the University of Tokyo. This work was supported by KAKENHI (Grant-in-Aid for Young Scientists (A) (No. 24689018) and on Innovative Areas "RNA regulation" (No. 23112702)) and the Global Center of Excellence Program for "Integrative Life Science Based on the Study of Biosignaling Mechanisms" from the Ministry of Education, Culture, Sports, Science and Technology of Japan, and the Cell Science Research Foundation.

REFERENCES

[1] Ambros V, Chen X. The regulation of genes and genomes by small RNAs. Development 2007;134:1635–41.

[2] Bartel DP. MicroRNAs: target recognition and regulatory functions. Cell 2009;136:215–33.

[3] Croce CM. Causes and consequences of microRNA dysregulation in cancer. Nat Rev Genet 2009;10:704–14.

[4] Suzuki HI, Miyazono K. Dynamics of microRNA biogenesis: crosstalk between p53 network and microRNA processing pathway. J Mol Med (Berl) 2010;88:1085–94.

[5] Matsuyama H, Suzuki HI, Nishimori H, Noguchi M, Yao T, Komatsu N, et al. miR-135b mediates NPM-ALK-driven oncogenicity and renders IL-17-producing immunophenotype to anaplastic large cell lymphoma. Blood 2011;118:6881–92.

[6] Kumar MS, Lu J, Mercer KL, Golub TR, Jacks T. Impaired microRNA processing enhances cellular transformation and tumorigenesis. Nat Genet 2007;39:673–7.

[7] Kumar MS, Pester RE, Chen CY, Lane K, Chin C, Lu J, et al. Dicer1 functions as a haploinsufficient tumor suppressor. Genes Dev 2009;23:2700–4.

[8] Lambertz I, Nittner D, Mestdagh P, Denecker G, Vandesompele J, Dyer MA, et al. Monoallelic but not biallelic loss of Dicer1 promotes tumorigenesis in vivo. Cell Death Differ 2010;17:633–41.

[9] Bahubeshi A, Tischkowitz M, Foulkes WD. miRNA processing and human cancer: DICER1 cuts the mustard. Sci Transl Med 2011;3:111ps146.

[10] Siomi H, Siomi MC. Posttranscriptional regulation of microRNA biogenesis in animals. Mol Cell 2010;38:323–32.

[11] Suzuki HI, Miyazono K. Emerging complexity of microRNA generation cascades. J Biochem 2011;149:15–25.

[12] Suzuki HI, Yamagata K, Sugimoto K, Iwamoto T, Kato S, Miyazono K. Modulation of microRNA processing by p53. Nature 2009;460:529–33.

[13] Suzuki HI, Arase M, Matsuyama H, Choi YL, Ueno T, Mano H, et al. MCPIP1 ribonuclease antagonizes dicer and terminates microRNA biogenesis through precursor microRNA degradation. Mol Cell 2011;44:424–36.

[14] Han J, Lee Y, Yeom KH, Nam JW, Heo I, Rhee JK, et al. Molecular basis for the recognition of primary microRNAs by the Drosha-DGCR8 complex. Cell 2006;125:887–901.

[15] Lund E, Guttinger S, Calado A, Dahlberg JE, Kutay U. Nuclear export of microRNA precursors. Science 2004;303:95–8.

[16] Ruby JG, Jan CH, Bartel DP. Intronic microRNA precursors that bypass Drosha processing. Nature 2007;448:83–6.

[17] Okamura K, Hagen JW, Duan H, Tyler DM, Lai EC. The mirtron pathway generates microRNA-class regulatory RNAs in Drosophila. Cell 2007;130:89–100.

[18] Cheloufi S, Dos Santos CO, Chong MM, Hannon GJ. A dicer-independent miRNA biogenesis pathway that requires Ago catalysis. Nature 2010;465:584–9.

[19] Cifuentes D, Xue H, Taylor DW, Patnode H, Mishima Y, Cheloufi S, et al. A novel miRNA processing pathway independent of Dicer requires Argonaute2 catalytic activity. Science 2010;328:1694–8.

[20] Gregory RI, Yan KP, Amuthan G, Chendrimada T, Doratotaj B, Cooch N, et al. The Microprocessor complex mediates the genesis of microRNAs. Nature 2004;432:235–40.

[21] Shiohama A, Sasaki T, Noda S, Minoshima S, Shimizu N. Nucleolar localization of DGCR8 and identification of eleven DGCR8-associated proteins. Exp Cell Res 2007;313:4196–207.

[22] Fuller-Pace FV. DExD/H box RNA helicases: multifunctional proteins with important roles in transcriptional regulation. Nucleic Acids Res 2006;34:4206–15.

[23] Tanner NK, Linder P. DExD/H box RNA helicases: from generic motors to specific dissociation functions. Mol Cell 2001;8:251–62.

[24] Fukuda T, Yamagata K, Fujiyama S, Matsumoto T, Koshida I, Yoshimura K, et al. DEAD-box RNA helicase subunits of the Drosha complex are required for processing of rRNA and a subset of microRNAs. Nat Cell Biol 2007;9:604–11.

[25] Sakamoto S, Aoki K, Higuchi T, Todaka H, Morisawa K, Tamaki N, et al. The NF90-NF45 complex functions as a negative regulator in the microRNA processing pathway. Mol Cell Biol 2009;29:3754–69.

[26] Michlewski G, Guil S, Semple CA, Caceres JF. Posttranscriptional regulation of miRNAs harboring conserved terminal loops. Mol Cell 2008;32:383–93.

[27] Kawahara Y, Mieda-Sato A. TDP-43 promotes microRNA biogenesis as a component of the Drosha and Dicer complexes. Proc Natl Acad Sci USA 2012;109:3347–52.

[28] Trabucchi M, Briata P, Garcia-Mayoral M, Haase AD, Filipowicz W, Ramos A, et al. The RNA-binding protein KSRP promotes the biogenesis of a subset of microRNAs. Nature 2009;459:1010–4.

[29] Nakamura T, Canaani E, Croce CM. Oncogenic All1 fusion proteins target Drosha-mediated microRNA processing. Proc Natl Acad Sci USA 2007;104:10980–5.

[30] Yu B, Bi L, Zheng B, Ji L, Chevalier D, Agarwal M, et al. The FHA domain proteins DAWDLE in Arabidopsis and SNIP1 in humans act in small RNA biogenesis. Proc Natl Acad Sci USA 2008;105:10073–8.

[31] Wu H, Sun S, Tu K, Gao Y, Xie B, Krainer AR, et al. A splicing-independent function of SF2/ASF in microRNA processing. Mol Cell 2010;38:67–77.

[32] Gruber JJ, Zatechka DS, Sabin LR, Yong J, Lum JJ, Kong M, et al. Ars2 links the nuclear cap-binding complex to RNA interference and cell proliferation. Cell 2009;138:328–39.

[33] Sabin LR, Zhou R, Gruber JJ, Lukinova N, Bambina S, Berman A, et al. Ars2 regulates both miRNA- and siRNA- dependent silencing and suppresses RNA virus infection in Drosophila. Cell 2009;138:340–51.

[34] Pawlicki JM, Steitz JA. Nuclear networking fashions pre-messenger RNA and primary microRNA transcripts for function. Trends Cell Biol 2010;20:52–61.

[35] Kim YK, Kim VN. Processing of intronic microRNAs. EMBO J 2007;26:775–83.

[36] Kataoka N, Fujita M, Ohno M. Functional association of the Microprocessor complex with the spliceosome. Mol Cell Biol 2009;29:3243–54.

[37] Morlando M, Ballarino M, Gromak N, Pagano F, Bozzoni I, Proudfoot NJ. Primary microRNA transcripts are processed co-transcriptionally. Nat Struct Mol Biol 2008;15:902–9.

[38] Ballarino M, Pagano F, Girardi E, Morlando M, Cacchiarelli D, Marchioni M, et al. Coupled RNA processing and transcription of intergenic primary microRNAs. Mol Cell Biol 2009;29:5632–8.

[39] Pawlicki JM, Steitz JA. Primary microRNA transcript retention at sites of transcription leads to enhanced microRNA production. J Cell Biol 2008;182:61–76.

[40] Pothof J, Verkaik NS, van IW, Wiemer EA, Ta VT, van der Horst GT, et al. MicroRNA-mediated gene silencing modulates the UV-induced DNA-damage response. EMBO J 2009;28:2090–9.

[41] Bates GJ, Nicol SM, Wilson BJ, Jacobs AM, Bourdon JC, Wardrop J, et al. The DEAD box protein p68: a novel transcriptional coactivator of the p53 tumour suppressor. EMBO J 2005;24:543–53.

[42] Watanabe M, Yanagisawa J, Kitagawa H, Takeyama K, Ogawa S, Arao Y, et al. A sub-family of RNA-binding DEAD-box proteins acts as an estrogen receptor alpha coactivator through the N-terminal activation domain (AF-1) with an RNA coactivator, SRA. EMBO J 2001;20:1341–52.

[43] Soussi T. p53 alterations in human cancer: more questions than answers. Oncogene 2007;26:2145–56.

[44] Song H, Xu Y. Gain of function of p53 cancer mutants in disrupting critical DNA dam-age response pathways. Cell Cycle 2007;6:1570–3.

[45] Adorno M, Cordenonsi M, Montagner M, Dupont S, Wong C, Hann B, et al. A Mutant-p53/Smad complex opposes p63 to empower TGFbeta-induced metastasis. Cell 2009;137:87–98.

[46] Song H, Hollstein M, Xu Y. p53 gain-of-function cancer mutants induce genetic instability by inactivating ATM. Nat Cell Biol 2007;9:573–80.

[47] Davis BN, Hilyard AC, Lagna G, Hata A. SMAD proteins control DROSHA-mediated microRNA maturation. Nature 2008;454:56–61.

[48] Bourguignon LY, Spevak CC, Wong G, Xia W, Gilad E. Hyaluronan-CD44 interac-tion with protein kinase C(epsilon) promotes oncogenic signaling by the stem cell marker Nanog and the Production of microRNA-21, leading to down-regulation of the tumor suppressor protein PDCD4, anti-apoptosis, and chemotherapy resistance in breast tumor cells. J Biol Chem 2009;284:26533–46.

[49] Yamagata K, Fujiyama S, Ito S, Ueda T, Murata T, Naitou M, et al. Maturation of microRNA is hormonally regulated by a nuclear receptor. Mol Cell 2009;36:340–7.

[50] Kawai S, Amano A. BRCA1 regulates microRNA biogenesis via the DROSHA microprocessor complex. J Cell Biol 2012;197:201–8.

[51] He L, He X, Lim LP, de Stanchina E, Xuan Z, Liang Y, et al. A microRNA component of the p53 tumour suppressor network. Nature 2007;447:1130–4.

[52] Braun CJ, Zhang X, Savelyeva I, Wolff S, Moll UM, Schepeler T, et al. p53-Responsive microRNAs 192 and 215 are capable of inducing cell cycle arrest. Cancer Res 2008;68:10094–104.

[53] Georges SA, Biery MC, Kim SY, Schelter JM, Guo J, Chang AN, et al. Coordinated regulation of cell cycle transcripts by p53-Inducible microRNAs, miR-192 and miR-215. Cancer Res 2008;68:10105–12.

[54] Yan HL, Xue G, Mei Q, Wang YZ, Ding FX, Liu MF, et al. Repression of the miR-17-92 cluster by p53 has an important function in hypoxia-induced apoptosis. EMBO J 2009;28:2719–32.

[55] Park SY, Lee JH, Ha M, Nam JW, Kim VN. miR-29 miRNAs activate p53 by targeting p85 alpha and CDC42. Nat Struct Mol Biol 2009;16:23–9.

[56] Brown CJ, Lain S, Verma CS, Fersht AR, Lane DP. Awakening guardian angels: drug-ging the p53 pathway. Nat Rev Cancer 2009;9:862–73.

[57] Fornari F, Gramantieri L, Giovannini C, Veronese A, Ferracin M, Sabbioni S, et al. MiR-122/cyclin G1 interaction modulates p53 activity and affects doxorubicin sensi-tivity of human hepatocarcinoma cells. Cancer Res 2009;69:5761–7.

[58] Voorhoeve PM, le Sage C, Schrier M, Gillis AJ, Stoop H, Nagel R, et al. A genetic screen implicates miRNA-372 and miRNA-373 as oncogenes in testicular germ cell tumors. Cell 2006;124:1169–81.

[59] Mudhasani R, Zhu Z, Hutvagner G, Eischen CM, Lyle S, Hall LL, et al. Loss of miRNA biogenesis induces p19Arf-p53 signaling and senescence in primary cells. J Cell Biol 2008;181:1055–63.

[60] Su X, Chakravarti D, Cho MS, Liu L, Gi YJ, Lin YL, et al. TAp63 suppresses metastasis through coordinate regulation of Dicer and miRNAs. Nature 2010;467:986–90.

[61] Zhang Z, Qin YW, Brewer G, Jing Q. MicroRNA degradation and turnover: regu-lating the regulators. Wiley Interdiscip Rev RNA 2012;3:593–600.

[62] Gantier MP, McCoy CE, Rusinova I, Saulep D, Wang D, Xu D, et al. Analysis of microRNA turnover in mammalian cells following Dicer1 ablation. Nucleic Acids Res 2011;39:5692–703.

[63] Diederichs S, Haber DA. Dual role for argonautes in microRNA processing and post-transcriptional regulation of microRNA expression. Cell 2007;131:1097–108.

[64] Hwang HW, Wentzel EA, Mendell JT. A hexanucleotide element directs microRNA nuclear import. Science 2007;315:97–100.

[65] Krol J, Busskamp V, Markiewicz I, Stadler MB, Ribi S, Richter J, et al. Characterizing light-regulated retinal microRNAs reveals rapid turnover as a common property of neuronal microRNAs. Cell 2010;141:618–31.

[66] Rissland OS, Hong SJ, Bartel DP. MicroRNA destabilization enables dynamic regulation of the miR-16 family in response to cell-cycle changes. Mol Cell 2011;43:993–1004.

[67] Ramachandran V, Chen X. Degradation of microRNAs by a family of exoribonucleases in Arabidopsis. Science 2008;321:1490–2.

[68] Chatterjee S, Grosshans H. Active turnover modulates mature microRNA activity in Caenorhabditis elegans. Nature 2009;461:546–9.

[69] Chatterjee S, Fasler M, Bussing I, Grosshans H. Target-mediated protection of endogenous microRNAs in C. elegans. Dev Cell 2011;20:388–96.

[70] Bail S, Swerdel M, Liu H, Jiao X, Goff LA, Hart RP, et al. Differential regulation of microRNA stability. RNA 2010;16:1032–9.

[71] Das SK, Sokhi UK, Bhutia SK, Azab B, Su ZZ, Sarkar D, et al. Human polynucleotide phosphorylase selectively and preferentially degrades microRNA-221 in human melanoma cells. Proc Natl Acad Sci USA 2010;107:11948–53.

[72] Umbach JL, Cullen BR. The role of RNAi and microRNAs in animal virus replication and antiviral immunity. Genes Dev 2009;23:1151–64.

[73] Moschos SA, Williams AE, Perry MM, Birrell MA, Belvisi MG, Lindsay MA. Expression profiling in vivo demonstrates rapid changes in lung microRNA levels following lipopolysaccharide-induced inflammation but not in the anti-inflammatory action of glucocorticoids. BMC Genomics 2007;8:240.

[74] O'Connell RM, Rao DS, Chaudhuri AA, Baltimore D. Physiological and pathological roles for microRNAs in the immune system. Nat Rev Immunol 2010;10:111–22.

[75] Ruggiero T, Trabucchi M, De Santa F, Zupo S, Harfe BD, McManus MT, et al. LPS induces KH-type splicing regulatory protein-dependent processing of microRNA-155 precursors in macrophages. FASEB J 2009;23:2898–908.

[76] Liang J, Song W, Tromp G, Kolattukudy PE, Fu M. Genome-wide survey and expression profiling of CCCH-zinc finger family reveals a functional module in macrophage activation. PLoS One 2008;3:e2880.

[77] Liang J, Wang J, Azfer A, Song W, Tromp G, Kolattukudy PE, et al. A novel CCCH-zinc finger protein family regulates proinflammatory activation of macrophages. J Biol Chem 2008;283:6337–46.

[78] Anantharaman V, Aravind L. The NYN domains: novel predicted RNAses with a PIN domain-like fold. RNA Biol 2006;3:18–27.

[79] Matsushita K, Takeuchi O, Standley DM, Kumagai Y, Kawagoe T, Miyake T, et al. Zc3h12a is an RNase essential for controlling immune responses by regulating mRNA decay. Nature 2009;458:1185–90.

[80] Han J, Pedersen JS, Kwon SC, Belair CD, Kim YK, Yeom KH, et al. Posttranscriptional crossregulation between Drosha and DGCR8. Cell 2009;136:75–84.

[81] Jura J, Skalniak L, Koj A. Monocyte chemotactic protein-1-induced protein-1 (MCPIP1) is a novel multifunctional modulator of inflammatory reactions. Biochim Biophys Acta 2012;1823:1905–13.

[82] Liang J, Saad Y, Lei T, Wang J, Qi D, Yang Q, et al. MCP-induced protein 1 deubiquitinates TRAF proteins and negatively regulates JNK and NF-kappaB signaling. J Exp Med 2010;207:2959–73.

[83] Gottwein E, Mukherjee N, Sachse C, Frenzel C, Majoros WH, Chi JT, et al. A viral microRNA functions as an orthologue of cellular miR-155. Nature 2007;450:1096–9.

[84] Gao G, Guo X, Goff SP. Inhibition of retroviral RNA production by ZAP, a CCCH-type zinc finger protein. Science 2002;297:1703–6.

[85] Sadler AJ, Williams BR. Interferon-inducible antiviral effectors. Nat Rev Immunol 2008;8:559–68.

[86] Barad O, Mann M, Chapnik E, Shenoy A, Blelloch R, Barkai N, et al. Efficiency and specificity in microRNA biogenesis. Nat Struct Mol Biol 2012;19:650–2.

[87] Bennasser Y, Chable-Bessia C, Triboulet R, Gibbings D, Gwizdek C, Dargemont C, et al. Competition for XPO5 binding between Dicer mRNA, pre-miRNA and viral RNA regulates human Dicer levels. Nat Struct Mol Biol 2011;18:323–7.

[88] Karube Y, Tanaka H, Osada H, Tomida S, Tatematsu Y, Yanagisawa K, et al. Reduced expression of Dicer associated with poor prognosis in lung cancer patients. Cancer Sci 2005;96:111–5.

[89] Merritt WM, Lin YG, Han LY, Kamat AA, Spannuth WA, Schmandt R, et al. Dicer, Drosha, and outcomes in patients with ovarian cancer. N Engl J Med 2008;359:2641–50.

[90] Martello G, Rosato A, Ferrari F, Manfrin A, Cordenonsi M, Dupont S, et al. A MicroRNA targeting dicer for metastasis control. Cell 2010;141:1195–207.

[91] Melo SA, Ropero S, Moutinho C, Aaltonen LA, Yamamoto H, Calin GA, et al. A TARBP2 mutation in human cancer impairs microRNA processing and DICER1 function. Nat Genet 2009;41:365–70.

[92] Melo SA, Moutinho C, Ropero S, Calin GA, Rossi S, Spizzo R, et al. A genetic defect in exportin-5 traps precursor microRNAs in the nucleus of cancer cells. Cancer Cell 2010;18:303–15.

[93] Hill DA, Ivanovich J, Priest JR, Gurnett CA, Dehner LP, Desruisseau D, et al. DICER1 mutations in familial pleuropulmonary blastoma. Science 2009;325:965.

[94] Heravi-Moussavi A, Anglesio MS, Cheng SW, Senz J, Yang W, Prentice L, et al. Recurrent somatic DICER1 mutations in nonepithelial ovarian cancers. N Engl J Med 2012;366:234–42.

[95] Melo S, Villanueva A, Moutinho C, Davalos V, Spizzo R, Ivan C, et al. Small molecule enoxacin is a cancer-specific growth inhibitor that acts by enhancing TAR RNA-binding protein 2-mediated microRNA processing. Proc Natl Acad Sci USA 2011;108:4394–9.

CHAPTER NINE

Nanoparticle-Based Delivery of siRNA and miRNA for Cancer Therapy

Rolando E. Yanes, Jie Lu, Fuyuhiko Tamanoi[1]

Department of Microbiology, Immunology and Molecular Genetics, Jonsson Comprehensive Cancer Center, University of California, Los Angeles, CA, USA
[1]Corresponding author: e-mail address: fuyut@microbio.ucla.edu

Contents

Abstract

Recent advance in the study of cancer genome has led to the realization that small interference RNAs (siRNAs) provide an attractive type of anticancer reagents that can inhibit the expression of genes that are upregulated or whose gene products are over-activated in human cancer. In addition, microRNAs (miRNAs) provide another type of anticancer agents that can affect multiple genes involved in signal transduction. However, the issue of delivery of these reagents needs to be overcome before they can be used extensively for cancer therapy. One of the solutions for this problem is to use nanoparticles that can carry siRNAs or miRNAs and deliver them to tumor. Various types of nanoparticles have been used to deliver these RNAs. They include mesoporous silica nanoparticles surface modified with cationic polymers as well as liposomes, dendrimers, or block copolymers. The use of siRNAs and miRNAs to block the signaling pathways important for cancer cells will be discussed. Different types of nanoparticles used to deliver these molecules will be mentioned and the prospect of these delivery vehicles playing an important role in the implementation of these reagents in medicine will be discussed.

The Enzymes, Volume 32
ISSN 1874-6047
http://dx.doi.org/10.1016/B978-0-12-404741-9.00009-X
185

1. INTRODUCTION

In the past years, significant progress has been made regarding cancer therapy. Elucidation of human genome sequences resulted in detailed understanding of molecular changes in human cancer [1,2]. Oncogenic mutations such as B-*RAF* mutations and K-*RAS* mutation as well as loss of tumor suppressors have been identified, providing us with a clearer understanding of signal transduction changes in cancer cells. This has led to the development of rational cancer therapy that targets these molecular changes [3]. Small molecule inhibitors have been identified against these molecular targets by screening of chemical compound libraries. Many of these belong to protein kinase inhibitor classes, but other classes of inhibitors are also being developed providing us with new arsenals to fight cancer.

Nucleic acid-based reagents have provided a promising source of anti-cancer reagents [4]. These include antisense oligonucleotides and siRNAs that can shut down gene expression. A variety of siRNAs have been developed and tested in preclinical models. In addition, a clinical study of siRNA delivered to melanoma has been published [5]. One of the advantages of using nucleic acid-based reagents is their specificity. Small molecule inhibitors could fit into proteins other than the target protein and cause non-specific effects while siRNAs are in general more specific for shutting down the expression of a particular gene. There have been some off-target effects observed with siRNAs but they are much lower than with chemical compounds. In addition, siRNAs can be designed so that they inhibit the expression of a mutant protein rather than the expression of the wild-type protein (discussed later).

Recently, it has been realized that another class of nucleic acid-based reagent, miRNA, can be developed as an anticancer reagent. In contrast to siRNAs, miRNAs affect the expression of multiple genes by either causing mRNA degradation or inhibiting translation [6]. Decreased expression of a particular miRNA in human cancer has been observed with a number of cases. In these instances, it is expected that the delivery of miRNA to those cancers will cause reversion of transformed phenotypes.

One of the major challenges regarding the use of siRNAs and miRNAs for cancer therapy is how to deliver them to tumor. While the RNA species can be used in tissue culture settings by using transfection reagents, this method of delivery is not applicable when administering *in vivo*. Viruses have been used to deliver siRNA (shRNA); however, there is a problem of having the virus jump into chromosomes causing the induction of unwanted gene expression. RNAs are also prone to nuclease attacks. Thus, modification of RNAs has been carried out to overcome this problem. The use of nanoparticles has the advantage of overcoming all these problems. Nanoparticles can provide protection to RNAs from degradation and enable targeting to tumor. This review is intended to survey the recent advance in the use of siRNAs, miRNAs, and their delivery using nanoparticles.

2. POTENTIAL USE OF siRNAS AND miRNAS FOR CANCER THERAPY

Human cancer is associated with the overactivation of a number of signaling pathways [7]. This includes the EGFR/Ras/Raf/MAPK signaling and insulin/PI3K/Akt/mTOR signaling. siRNAs provide powerful agents to shut down the expression of genes that control signaling pathways. An interesting application of the use of siRNA is to specifically shut down the expression of a gene that is mutated in cancer. siRNAs against a mutant form of Ras is an example of this application. A number of human cancer cases have been associated with the loss of a particular miRNA species. In this case, reintroduction of miRNA is expected to reverse the transformed phenotypes. In this section, some examples of these RNAs will be discussed.

2.1. siRNA

A variety of siRNAs have been used to silence signaling pathways over-activated in human cancer. For example, receptor tyrosine kinases have been shown to be overexpressed or overactivated in some types of cancer. Targeting these receptors is one approach that can decrease cancer cell proliferation. Chen and coworkers developed a system for stable expression of siRNA against *TEL-PDGFβR* oncogene and demonstrated that there was a 90% decrease in *TEL-PDGFβR* expression and its downstream effectors [8]. This resulted in a significant attenuation in proliferation of *TEL-PDGFβR* transformed Ba/F3 cells. Furthermore, when mice were injected with *TEL-PDGFβR* transformed Ba/F3 cells expressing TEL–PDGFβR siRNA, they had a longer survival time than the mice injected with cells not expressing the siRNA. Another receptor that has been targeted in a similar approach is EphA2, which plays a role in ovarian, breast, and pancreatic cancer. Silencing of the *EphA2* gene with siRNA delivered by DOPC liposomes resulted in the reduction of tumor growth in mice [9].

Bcr-abl is an oncogene that is produced by chromosome translocation seen in chronic myelogenous leukemia (CML). A number of small molecule drugs have been developed providing support for the idea that the inhibition of this gene is critical for this type of leukemia. siRNAs targeting this fusion oncogene have been developed. Scherr and coworkers used reporter gene constructs to select siRNAs that displayed efficient gene knockdown for *bcr-abl* mRNA [10]. They were able to develop siRNAs that achieved up to 87% gene silencing in *bcr-abl* positive cell lines and in primary cells from CML patients. The specificity of the siRNA for the *bcr-abl* fusion oncogene was high enough that the *c-abl* and *c-bcr* mRNA levels remained unaffected in the cells. The anti-*bcr-abl* siRNA inhibited BCR-ABL dependent proliferation in a *bcr-abl*-positive cell line demonstrating that silencing this gene could be beneficial to combating cancers originated through this translocation. CML is often treated with Imatinib, a selective tyrosine kinase inhibitor that has proved to be efficient in controlling leukemic growth. However, there are leukemic cells that are resistant to this treatment making imatinib insufficient to eradicate the disease. Wohlbold *et al.* demonstrated that knocking down BCT-ABL protein expression with siRNA increases sensitivity to imatinib in a cell line expressing the imatinib-resistant BCR-ABL kinase domain mutation His396Pro [11]. Furthermore, Koldehoff *et al.* performed a clinical study in which he injected a patient suffering from imatinib-resistant CML with liposomes carrying anti-*bcr-abl* siRNA [12]. He observed inhibition of the overexpressed *bcr-abl* oncogene resulting in an increased apoptosis of CML cells. This suggests that targeting the *bcr-abl* oncogene with siRNA could be an efficient therapy for patients suffering CML.

B-*Raf* is mutated in more than 50% of melanoma making this oncogene an attractive target for anticancer therapy. *Skp-2* is another gene linked to different types of cancer as its overexpression leads to enhanced degradation of p27^{Kip1}, which regulates G1/S transition of the cell cycle and inhibits Rho A signaling. This results in deregulation of cell growth and cell motility. Sumimoto and coworkers used siRNA against B-*Raf* and *Skp-2* and showed that suppression of these oncogenes results in the inhibition of *in vitro* cell growth and invasive ability of melanoma cells [13].

Human cancer cells have changes that contribute to increased survival by preventing apoptosis. One of the genes blocking apoptosis is Bcl-2, which prevents activation of the extrinsic death pathway. This pathway can be activated by a number of extrinsic apoptosis-inducing agents, such as tumor necrosis factor-related apoptosis-inducing ligand (TRAIL). Kock *et al.*

combined the knockdown of *Bcl-2* with shRNA with the expression of a secretable form of TRAIL (S-TRAIL) and demonstrated this treatment results in accelerated S-TRAIL mediated apoptosis in glioma cells [14]. Furthermore, the combined effect of this treatment resulted in a significant reduction of gliomas in a human glioma model expressing EGFRvIII.

Human telomerase reverse transcriptase (hTERT) continuously synthesizes telomeric DNA sequences at the ends of chromosomes. These sequences preserve chromosomal integrity during mitosis. While in normal cells telomerase activity is only detected in cells with proliferative potential, such as germ line and hematopoietic cells, the majority of cancer cells exhibit telomerase activity. Ge *et al.* demonstrated that RNAi targeting *hTERT* expression induces apoptosis and inhibits the proliferation of lung cancer cells [15]. This study provides an experimental evidence that *hTERT* is a good target for RNAi mediated cancer cell death induction.

An interesting example of siRNA has been developed that targets the *RAS* oncogene. Mutations of the *RAS* gene are observed in more than 90%, 50%, and 30% of pancreatic cancer, lung cancer, and colon cancer, respectively. While there are three isoforms, K-*RAS*, N-*RAS*, and H-*RAS*, most mutations occur on K-*RAS*. The mutations are single amino acid changes in the residues that are critical for GTPase activity of Ras. Thus, the mutations lead to active Ras that is bound with GTP. Brummelkamp *et al.* developed a vector to specifically inhibit expression of a mutant form of K-*RAS* (K-*RAS*^{V12}) through siRNA. Silencing K-*RAS*^{V12} resulted in loss of anchorage-independent growth and tumorigenicity of the pancreatic cancer cell line CAPAN-1 [16]. In a similar study, Zhu *et al.* demonstrated that siRNAs targeting a different mutant K-*RAS* can significantly decrease the expression of this mutant gene [17]. Reduction of K-*RAS* expression in PC-7 and PANC-1 pancreatic cell lines leads to inhibition of cell growth and intratumor injection of the siRNA encapsulated in polyethyleneimine in mice inhibits tumor growth.

2.2. miRNA

Recent studies have shown that human cancer is associated with changes in miRNA profile [18]. In fact, the use of miRNA profile as novel biomarkers for cancer has been proposed. An example is the use of fecal miRNAs as biomarkers for pancreatic cancer [19]. This study demonstrated that the feces of patients with pancreatic cancer have a lower concentration of miR-216a, -196a, -143, and -155 when compared with controls. This suggests that fecal miRNA biomarkers could be used as a tool for pancreatic cancer screening research. Changes in miRNA expression are associated with cancer. In some types of cancer, increased levels of miRNA can be detected while in other types there is a significant decrease in miRNA expression. In the latter case, reintroduction of miRNA is an approach that can be used for therapy.

A variety of microRNAs have exhibited the ability to reverse transformed phenotypes. In particular, epithelial to mesenchymal transition (EMT) implicated in cancer metastasis was found to be reversed by reintroducing a number of miRNAs. For example, endometrial cancer (EC) cells have been found to express B lymphoma mouse Moloney leukemia virus insertion region 1 (BMI-1), which induces EMT in cancer cells. Ectopic expression of miRNA-194 in these cells was shown to inhibit EMT by downregulating BMI-1 via direct binding to the BMI-1 3′-untranslated region [20]. Furthermore, miRNA-194 induced mesenchymal to epithelial transition (MET) in EC cells by restoring E-cadherin, reducing Vimentin expression and inhibiting cell invasion *in vitro*. Similarly, miRNA-145 suppresses the invasion–metastasis cascade in gastric cancer (GC) by inhibiting *N*-cadherin protein translation and indirectly downregulating matrix metalloproteinase 9. This miRNA has been found to be downregulated in human GCs and GC cell lines and ectopic reconstitution of this miRNA suppresses cell migration and invasion *in vitro* and *in vivo* [21]. In another study in GC, it was found that miR-7 is downregulated in GC cell lines and is inversely correlated with GC metastasis [22]. One of the targets of miR-7 is the insulin-like growth factor-1 receptor (IGF1R) oncogene, which regulates cell growth and tumor invasion. Transfection of miR-7 into GC cells increased E-cadherin expression and partially reversed the EMT phenotype.

In the case of breast cancer, it was shown that the loss of miR-133a expression is associated with metastasis, high clinical stages and shorter relapse-free survival. The target of this miRNA is Fascin1 (FSCN1), an actin-bundling protein that is normally expressed in mesenchymal, neuronal, and endothelial cells but not in normal epithelial cells. Restoration of miR-133a inhibited breast cancer cell growth and invasion [23]. Similarly, in pancreatic cancer cells, miR-146a has been shown to be downregulated when compared with normal human pancreatic duct epithelial cells. This miRNA suppresses the invasion of pancreatic cancer cells by decreasing the expression of EGFR, MTA-2, IRAK-1, and NF-κB [24]. miR-126 is another microRNA shown to act as tumor suppressor via the regulation of ADAM9 (disintegrin and metalloproteinase domain-containing protein 9). Reexpression of miR-126 leads to reduced cellular migration, invasion, and induction of epithelial marker E-cadherin [25]. One of the factors contributing to tumor invasion and metastasis in pancreatic cancer is the deregulation of miR-200, which regulates the expression of membrane type-1 matrix metalloproteinase (MT1-MMP) and PTEN. In chemoresistant pancreatic cancer cell lines, the miR-200 family is not expressed. Induction of miR-200c reexpression by two novel agents, CDF and BR-DIM, has been shown to result in MT1-MMP downregulation and PTEN reexpression of miR-200 leading to reduced tumor cell aggressiveness [26].

3. NANOPARTICLE-BASED DELIVERY

siRNA is a powerful tool for gene therapy with great potential for biomedical applications, providing precise selectivity of targeting therapeutics. However, siRNA molecules cannot simply cross the cell plasma membrane, requiring an effective carrier system for their delivery. Many methods to deliver siRNA into human cells were attempted, including viral and nonviral vectors (Fig. 9.1). However, numerous studies have shown that viral vectors are associated with systemic toxicity and immunogenicity. Problems such as poor intracellular uptake, instability, and nonspecific immune stimulation need to be overcome for successful and practical nonviral siRNA delivery. Therefore, developing a more stable, highly effective siRNA delivery system with optimal cellular integration and low toxicity is of utmost importance. Recent advancement of nanomaterials and nanoscience provides the possibilities for a novel nanoparticle-based siRNA delivery vehicle.

3.1. Mesoporous silica-based approach

Mesoporous silica nanoparticles (MSNs), which are nanometer-sized drug carriers, have been proved to be a versatile delivery vehicle with attractive features such as stability, large surface area, tunable pore sizes and volumes, and easy encapsulation of drugs, proteins, and biogenic molecules. The synthesis of MSNs is simple, low cost, and scalable. These nanoparticles have a good biocompatibility. They are generally recognized as safe by the FDA, as this material has been widely used in cosmetics as well as food additives. MSNs represent a novel delivery vehicle to cells that are able to carry small molecule compounds and nucleic acids to cells (Fig. 9.1). For increasing the loading capacity of siRNA, the well-established chemistry can be employed to modify the surface of MSNs such as modification with polyamidoamine [28] and mannosylated polyethylenimine [29]. The positive charge on the surface increases the electrostatic interaction between MSNs and the

Viral infection	Chemical transfection	Mesoporous silica nanoparticles (MSNs)	Large pore MSNs	Organic/inorganic silica hybrid nanoparticles
Nucleic acids (DNA, siRNA, ASO)	Nucleic acids (DNA, siRNA, ASO)	Nucleic acids (DNA, siRNA, ASO)	Nucleic acids (DNA, siRNA, ASO)	Nucleic acids (DNA, siRNA, ASO)
		Chemotherapeutic compounds		Chemotherapeutic compounds

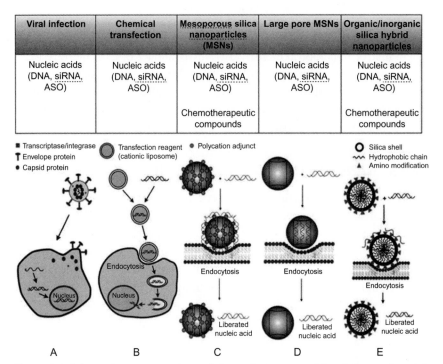

Figure 9.1 Schematic representations of the techniques used for delivery of nucleic acid-based reagents into plant and mammalian cells. Nucleic acid molecules theoretically bind to positively charged chemical modifications on the external surface of MSNs and hybrid nanoparticles. *Reprinted with permission from Hom* et al. *[27].*

negatively charged siRNAs and also helps escape of MSNs from endosome by proton sponge effect. MSNs can also be modified with a variety of surface functional groups to improve their biocompatibility and design for different purposes, such as light–responsive drug delivery nanoimpeller, pH–responsive nanovalve, intracellular reductase-responsive nanosnap top nanomachines, as well as positive cancer-targeting drug delivery system [30]. Hybrids of silica and other organic/inorganic materials have also been used in various applications including enzyme encapsulation and plasmid DNA transfection [31,32]. Furthermore, studies have shown that MSNs preferentially accumulate in tumors due to the enhanced permeability and retention (EPR) effect of tumors, and the accumulation can be enhanced by positive targeting with tumor-specific ligand on the surface [33]. Doses of MSNs nontoxic and biocompatible to animals can be achieved [31,34].

A number of groups have successfully shown that siRNA can be effectively delivered into cells by MSNs, and knocking down of both exogenous and endogenous gene expression was observed. Hom *et al.* [35] showed that coating with cationic polymer PEI to MSNs (Fig. 9.2) dramatically improved the binding of siRNA and its release, compared to siRNA/PEI complex. Importantly, siRNA after binding to MSNs was much more resistant to RNase digestion, suggesting that the binding to MSN provided protection of siRNA from enzymatic degradation which is critical for *in vivo* siRNA delivery applications. Delivery of siRNA against GFP resulted in the decrease of fluorescence of cells expressing GFP (Fig. 9.3). Compared to other nonviral transfection vehicles, such as Lipofectamine 2000, PEI–MSNs showed much less cytotoxicity to cells. In this study, efficacy to shut down signal transduction in human cancer cells was demonstrated by delivering siRNA against K-ras resulting in the inhibition of Erk phosphorylation in a pancreatic cancer cell line. Furthermore, delivery of siRNA against Akt by the PEI-MSN resulted in the inhibition of expression of this key protein kinase, which is activated in a wide range of human cancers.

In addition to the simple delivery of siRNA, Chen *et al.* showed that utilizing MSNs to simultaneously deliver anticancer drug and siRNA into multidrug resistant cancer cells for enhanced efficacy of chemotherapy [36]. By loading anticancer drugs in the pores and siRNA absorbed on the surface, the Bcl-2 siRNA delivered by MSNs was shown to effectively silence the Bcl-2 mRNA and substantially enhance the anticancer efficacy of Doxurubicin, proving the versatile capacity of MSNs as both drug and siRNA delivery vehicles for efficient cancer therapy. Delivery of siRNA against multidrug resistance (MDR) gene together with the delivery of Doxorubicin was reported by Meng *et al.* [37]. On the other hand, instead of absorbing siRNA on the surface of MSNs, Li *et al.* have successfully

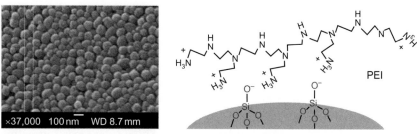

Figure 9.2 SEM pictures of MSNs on the left panel and schematic representation of PEI attachment on the surface of MSNs on the right panel. (For color version of this figure, the reader is referred to the online version of this chapter.).

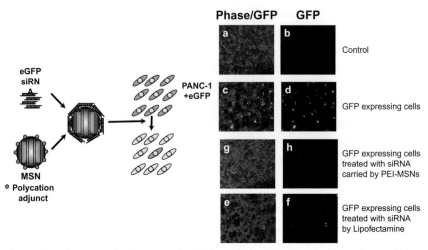

Figure 9.3 Conceptual diagram of siRNA-mediated EGFP gene silencing utilizing mesoporous silica nanoparticles as delivery vehicles. Polycation adjuncts (i.e., PEI) allow binding between nucleic acids and MSNs (Left). Knockdown of GFP (green fluorecent protein) expression by siRNA delivered by PEI–MSNs (Right). *Reprinted (adapted) with permission from Hom et al. [35].* (See color plate section in the back of the book.)

packaged siRNA into the mesopores of MSNs under a strongly dehydrated solution condition [38]. The pores provided excellent protection effect of siRNA, and the release siRNA in the cytoplasm resulted in the knockdown of both exogenous enhanced green fluorescent protein (EGFP) gene and endogenous B-cell lymphoma 2 (Bcl-2) gene. Larger-sized pore hollow silica nanoparticles were also synthesized and showed increased siRNA-loading capabilities inside of the pores and intracellular transfection efficiencies [39,40].

Using animal models, Tarratula *et al.* showed that local delivery of MSN by inhalation resulted in preferential accumulation in the mouse lungs, limiting their accumulation in other organs. Delivery of both anticancer drugs (doxorubicin and cisplatin) and siRNA against MRP1 and BCL2 mRNA to suppress pump and nonpump cellular resistance was achieved in non-small cell lung cancers [41]. Cancer-targeting delivery of specific siRNA was also demonstrated by conjugating specific peptide SP94 [42] and LHRH [22] for lung cancers on the surface of MSNs.

All these pioneering studies suggested that MSNs possess several attractive characteristics for use in the delivery of siRNA. siRNA can be loaded, protected, and delivered into human cells by MSNs. Precise cancer-targeting delivery of siRNA can be achieved by modifying the surface of MSNs. Therefore, MSNs provide an attractive platform for siRNA delivery and gene therapy for a wide range of diseases. However, more preclinical and clinical studies need to be done before MSNs-moderated siRNA delivery can be used for gene therapy in clinical settings.

3.2. Liposome-based approach

Liposomes are spherical structures composed of a lipid bilayer and an aqueous core where molecules can be entrapped. They are optimal for siRNA therapy because they can fuse with the cell membrane and deliver the cargo into the cytoplasm where the siRNA can inhibit mRNA translation. Liposomes can improve the serum stability, solubility, circulation half-life, biodistribution, and target tissue specificity of siRNA. Some siRNA delivery systems consist of multiple lipid components that include cationic lipids, fusogenic helper lipids, and polyethylene glycol (PEG)-lipids. Cationic lipids encourage inter-action between the lipid bilayer and the negatively charged siRNA, which increases the nucleic acid encapsulation efficacy reaching over 95% efficacy when using modern techniques [43,44]. These lipids also provide the liposomes with a net positive charge that enables binding to anionic cell surface molecules, such as sulfated proteoglycans and sialic acids, increasing cellular uptake [45]. Fusogenic helper lipids facilitate the intracellular delivery of complexed nucleic acid drugs by fusing with the membranes of the target cells. Upon fusion with either the plasma or endosomal membranes, the liposomes deliver their contents directly into the cytoplasm. Fusogenic lipids promote destabilization of the membranes by inducing the reverse hexagonal H_{II} phase, which do not support bilayer structures. The degree of lipid saturation determines the lipid's affinity for the fusogenic H_{II} phase [46]. Some cationic lipids can function in the absence of helper fusogenic lipids, either alone or in the presence of nonfusogenic lipid cholesterol [47]. Fusogenic formulations are more likely to interact with the vascular endothelium, blood vessels, and other nontarget systems. That's why PEG is used to transiently shield the fusogenic potential of the system. PEG-lipid conjugates are incorporated in liposomal nucleic acid formulations to stabilize the nascent particle and to prevent aggregation. PEG conjugates form a hydrophilic layer that shields the hydrophobic lipid layer, thus preventing the association of serum proteins and reducing the uptake by the reticuloendothelial system. The formulation of liposomal components is an area of continuous innovation in an attempt to improve the stability, circulation half-life, and biodistribution of the delivery system.

Advances in liposome formulation have led to the development of siRNA delivery systems that demonstrate efficient gene silencing (Fig. 9.4). For example, liposomes consisting dimethyl-hydroxyethyl-aminopropane-carbamoyl-cholesterol and dioleoylphosphatidylethanolamine in equimolar proportions are capable of delivering siRNA molecules resulting in over 90% knockdown of VEGF in the cancer cell lines A431 and MDA–MB 231 [49]. This is important as VEGF promotes angiogenesis in tumors and decreasing its production could inhibit tumor growth. Other liposome formulations have led to the development of targeted siRNA delivery. Liposomes with the F3 peptide on their surface, which is internalized by nucleolin, a receptor overexpressed on the surface of cancer cells and endothelial cells of

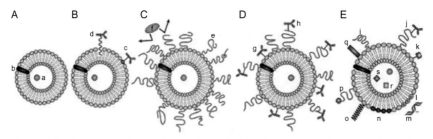

Figure 9.4 (A) Early traditional phospholipids "plain" liposomes with water soluble drug (a) entrapped into the aqueous liposome interior and water-insoluble drug (b) incorporated into the liposomal membrane (these designations are not repeated on other figures). (B) Antibody-targeted immunoliposome with antibody covalently coupled (c) to the reactive phospholipids in the membrane, or hydrophobically anchored (d) into the liposomal membrane after preliminary modification with a hydrophobic moiety. (C) Long-circulating liposome grafted with a protective polymer (e) such as PEG, which shields the liposome surface from the interaction with opsonizing proteins (f). (D) Long-circulating immunoliposome simultaneously bearing both protective polymer and antibody, which can be attached to the liposome surface (g) or, preferably, to the distal end of the grafted polymeric chain (h). (E) New generation liposome, the surface of which can be modified (separately or simultaneously) by different ways. Among these modifications are: the attachment of protective polymer (i) or protective polymer and targeting ligand, such as antibody (j); the attachment/incorporation of the diagnostic label (k); the incorporation of positively charged lipids (l) allowing for the complexation with DNA (m); the incorporation of stimuli-sensitive lipids (n); the attachment of stimuli-sensitive polymer (o); the attachment of cell-penetrating peptide (p); the incorporation of viral components (q). In addition to a drug, liposome can loaded with magnetic particles (r) for magnetic targeting and/or with colloidal gold or silver particles (s) for electron microscopy. *Reprinted with permissions from Torchilin et al. [48].* (See color plate section in the back of the book.).

tumor blood vessels have been developed [50]. This system can specifically deliver siRNA to the breast cancer cells with limited effect on human fibroblasts. In another study, liposomes with a different formulation were targeted to the non-small cell lung cancer cell line NCI-NH322 via anti-EGFR antibodies [51]. These liposomes were able to silence genes in NCI-NH322 cells but not in NIH-3T3 cells, which do not have the EGF receptor.

3.3. Polymer-based approach

Polymer nanoparticles with a micelle structure have also been used for siRNA delivery with promising results. Nanoparticles used for nucleic acid delivery usually consist of a cationic polymer that associates with the negatively charged phosphate backbone of nucleic acids and poly-(ethylene glycol) (PEG) which increases the stability [52,53] and keeps the nanoparticles soluble in aqueous solutions [54]. There are many cationic polymers that can be used to produce micellar nanoparticles and the physiochemical characteristics of each polymer affect the nanoparticle's stability, biodistribution, and cellular uptake mechanism. Given the vast diversity of nanoparticles, we would like to focus on cyclodextrin polymer (CDP) nanoparticles developed by Davis and coworkers and which have entered phase I clinical trials.

In 1999, Davis and coworkers used a CDP for the delivery of pDNA [55], but the discovery that siRNAs could be used for gene silencing led them to shift their focus to siRNA delivery. Modifications to the CDP formulation resulted in the development of a copolymer nanoparticle consisting of three components: CDP, adamantane-PEG (AD-PEG), and adamantane-PEG-Transferrin (AD-PEG-Tf) [56] (Fig. 9.5). The CDP is a short polycation that consists of five monomers and is capped with imidazole groups to facilitate escape from the endocytic vesicle [58]. The CDP is the one that binds and condenses the siRNA molecules while AD-PEG is used to stabilize the CDP-siRNA nanoparticles and prevent protein induced aggregation in serum. AD-PEG interacts with the surface accessible cyclodextrins on the surface of the CDP-siRNA nanoparticles. The AD-PEG-Tf component confers targeting functionality to the nanoparticles as it allows for a multivalent binding to upregulated copies of transferrin receptor on the cell surface and enhances uptake. *In vivo* studies with murine tumor models demonstrated that these nanoparticles were able to deliver siRNA against ribonucleotide reductase subunit 2 (RRM2) resulting in silencing of this gene [59].

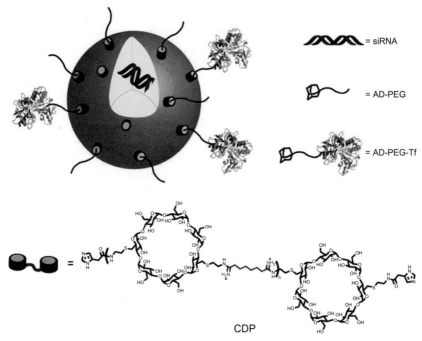

Figure 9.5 A graphical representation of the CDP delivery system. *Reprinted with permission from Alabi et al. [57].* (See color plate section in the back of the book.)

Further modifications to the cyclodextrin nanoparticles resulted in the development of CALAA-01 for clinical trials. These nanoparticles have a diameter of approximately 70 nm and contain 10,000 CDP molecules, 2000 siRNA molecules, 4000 AD-PEG, and 100 AD-PEG-Tf [56]. Studies in nonhuman primates demonstrated that these particles are well tolerated at doses up to 27 mg siRNA/kg and they have a translated efficacy in the range of 0.6–12 mg siRNA/kg in monkeys [57]. The phase I clinical trial consisted in administering CALAA-01 containing RMM2 siRNA to human melanoma patients [5]. This trial resulted in the reduction of both RMM2 mRNA and RMM2 protein levels in tumor tissue. These were the first micellar nanoparticles to show siRNA delivery in humans and mechanistic evidence of RNAi in humans.

3.4. Other nanoparticles

Quantum dots have received a lot of attention because of their fluorescence properties that can be used for imaging, but a new application of these particles is delivery of siRNA. These nanoparticles are surface modified to allow siRNA binding and significant gene silencing has been achieved in a number of studies. Luminescence silicon quantum dots (siQDs) with 2-vinylpyridine covalently attached to their surface have been used to deliver ABCB1 siRNA to Caco-2 cells [60]. The 2-vinylpyridine binds the siRNA electrostatically and once the quantum dots are taken up by cells, the siRNA is released in the cytoplasm resulting in downregulation of P-glycoprotein. In another study, CdSe/ZnSe quantum dots were surface modified with L-arginine or β-cyclodextrin (β-CD)-L-arginine to deliver siRNA against the HPV18 E6 gene in HeLa cells [61]. Both of these QDs showed low cytotoxicity, protection of siRNA, efficient siRNA delivery, and gene silencing, but the β-CD-L-arginine QDs were superior in all aspects. In a follow-up study, the authors used the β-CD-L-arginine QDs to co-deliver siRNA targeting the multidrug resistance (MDR1) gene and doxorubicin (Dox) [62]. They were able to decrease MDR1 expression and improve the effect of Dox on HeLa cells resulting in higher cytotoxicity of the drug.

There are other types of nanoparticles that have been used for siRNA delivery with promising results. For example, Lin *et al.* developed a high-density lipoprotein-mimicking peptide-phospholipid scaffold (HPPS) for siRNA delivery [63]. This delivery system was able to deliver siRNA directly to the cytosol bypassing endosomal trapping. The siRNA was cholesterol-modified in order to be loaded onto the nanoparticles. The HPPS carrying siRNA targeting the *BCL-2* gene was injected into mice bearing scavenger receptor class B type 1 overexpressed tumors. HPPS prolonged the blood circulation of the siRNA, improved its biodistribution, and facilitated uptake by the tumor. This resulted in downregulation of the Bcl-2 protein in the cancer cells, induction of apoptosis and inhibition of tumor growth.

4. SUMMARY AND FUTURE PROSPECTS

The use of siRNA and miRNA for cancer therapy is moving toward clinical application due to the increased knowledge in gene expression deregulation that contributes to cancer, as well as novel delivery vehicles that could make *in vivo* application possible. As seen in this review, there are many oncogenes and signaling pathways that have been targeted using siRNA or miRNA with promising results. In some instances, the silencing of oncogenes has led to cancer cell death or decrease of tumor growth. Even though there were promising results in terms of siRNA and miRNA development for specific oncogenes and efficient expression knockdown *in vitro*, there was a big hurdle that needed to be overcome in order to implement this type of treatment in clinics. Nucleic acid molecules do not exhibit good biodistribution, are not readily taken up by cells, and have a low half-life *in vivo*. Therefore, they needed to be packaged into a delivery vehicle in order to achieve efficient gene knockdown *in vivo*. Viruses were one of the first delivery vehicles for nucleic acid-based therapy but the immunogenicity of the virus limited their efficacy. Advances in nanotechnology resulted in an array of new organic and inorganic nanoparticles that have shown efficient siRNA delivery, to the point that clinical trials are being carried out to determine if they could be used in clinics in the near future.

ACKNOWLEDGMENT

This work is supported by NIH Grant CA133697 and the NIH Biotechnology Training Grant T32GM067555.

REFERENCES

[1] Deepak N, Zack TI, Ren Y, Strickland MR, Lamothe R, Schumacher SE, et al. Cancer vulnerabilities unveiled by genomic loss. Cell 2012;150:842–54.
[2] The Cancer Genome Atlas Research Network. Comprehensive genomic characterization defines human glioblastoma genes and core pathways. Nature 2008;455:1061–8.
[3] Zhang Z, Li M, Rayburn ER, Hill DL, Zhang R, Wang H, et al. Oncogenes as novel targets for cancer therapy (part I): growth factors and protein tyrosine kinases. Am J Pharmacogenomics 2005;5:173–90.
[4] Seyhan AA. RNAi: a potential new class of therapeutic for human genetic disease. Hum Genet 2011;130:583–605.
[5] Davis ME, Zuckerman JE, Choi CH, Seligson D, Tolcher A, Alabi CA, et al. Evidence of RNAi in humans from systematically administered siRNA via targeted nanoparticles. Nature 2010;464:1067–70.
[6] Fabian MR, Sonenberg N. The mechanics of miRNA-mediated gene silencing: a look under the hood of miRISC. Nat Struct Mol Biol 2012;19:586–93.
[7] Hanahan D, Weinberg RA. Hallmarks of cancer: the next generation. Cell 2011;144:646–74.
[8] Chen J, Wall NR, Kocher K, Duclos N, Fabbro D, Neuberg C, et al. Stable expression of small interfering RNA sintesizes TEL-PDGFβR to inhibition with imatinib or rapamycin. J Clin Invest 2004;113:1784–91.
[9] Landen Jr. CN, Chavez-Reyes A, Bucana C, Schmandt R, Deavers MT, Lopez-Berestein G, et al. Therapeutic *EphA2* gene targeting *in vivo* using neutral liposomal small interfering RNA delivery. Cancer Res 2005;65:6910–8.
[10] Scherr M, Battmer K, Winkler T, Heindenreich O, Ganser A, Eder M. Specific inhibition of bcr-abl gene expression by small interfering RNA. Blood 2003;101:1566–9.
[11] Wohbold L, Van Der Kuip H, Meithing C, Vornlocher H-P, Knabbe C, Duyster J, et al. Inhibition of bcr-abl gene expression by small interfering RNA sensitizes for imatinib mysylate (STI571). Blood 2003;102:2236–9.
[12] Koldehoff M, Steckel NK, Beelen DW, Elmaagacli AH. Therapeutic application of small interfering RNA directed against bcr-abl transcripts to a patient with imatinib-resistant chronic myeloid leukemia. Clin Exp Med 2007;7:47–55.

[13] Sumimoto H, Hirata K, Yamagata S, Miyoshi H, Miyagishi M, Taira K, et al. Effective inhibition of cell growth and invasion of melanoma by combined suppression of BRAF (V599E) and Skp2 with lentiviral RNAi. Int J Cancer 2006;118:472–6.

[14] Kock N, Kasmieh R, Weissleder R, Shah K. Tumor therapy mediated by lentiviral expression of shBcl-2 ans S-TRAIL. Neoplasia 2007;9:435–42.

[15] Ge L, Shao W, Zhang Y, Qiu Y, Cui D, Huang D, et al. RNAi targeting of hTERT gene expression induces apoptosis and inhibits the proliferation of lung cancer cells. Oncol Lett 2011;2:1121–9.

[16] Brummelkamp TR, Bernards R, Agami R. Stable suppression of tumorigenicity by virus-mediated RNA interference. Cancer Cell 2002;2:243–7.

[17] Zhu H, Liang ZY, Ren XY, Liu TH. Small interfering RNAs targeting mutant K-ras inhibit human pancreatic carcinoma cells growth in vitro and in vivo. Cancer Biol Ther 2006;5:1693–8.

[18] Liu X, Liu L, Xu Q, Wu P, Zuo X, Ji A. MicroRNA as a novel drug target for cancer therapy. Expert Opin Biol Ther 2012;12:573–80.

[19] Link A, Becker V, Goel A, Wex T, Malfertheiner P. Feasibility of fecal microRNAs as novel biomarkers for pancreatic cancer. PLoS One 2012;7:e42933.

[20] Dong P, Kaneuchi M, Watari H, Hamada J, Sudo S, Ju J, et al. MicroRNA-194 inhibits epithelial to mesenchymal transition of endometrial cancer cells by targeting oncogene BMI-1. Mol Cancer 2011;10:99.

[21] Gao P, Xing A-Y, Zhou G-Y, Zhang T-G, Zhang J-P, Gao C, et al. The molecular mechanism of microRNA-145 to suppress invasion-metastasis cascade in gastric cancer. Oncogene 2012; http://dx.doi.org/10.1038/onc.2012.61.

[22] Zhao X, Dou W, He L, Liang S, Tie J, Liu C, et al. MicroRNA-7 functions as an anti-metastatic microRNA in gastric cancer by targeting insulin-like growth factor-1 receptor. Oncogene 2012;1–10. http://dx.doi.org/10.1038/onc.2012.156.

[23] Wu Z-S, Wang C-Q, Xiang R, Liu X, Ye S, Yang X-Q, et al. Loss of miR-133a expression associated with poor survival of breast cancer and restoration of miR-133a expression inhibited breast cancer cell growth and invasion. BMC Cancer 2012;12:51.

[24] Li Y, VandenBoom II TG, Wang Z, Kong D, Ali S, Philip PA, et al. miRNA-146a suppresses invasion of pancreatic cancer cells. Cancer Res 2010;70:1486–95.

[25] Hamada S, Satoh K, Fujibuchi W, Hirota M, Kanno A, Unno J, et al. MiR-126 acts as a tumor suppressor in pancreatic cáncer cells via the regulation of ADAM9. Mol Cancer Res 2011;10:3–10.

[26] Soubani O, Ali AS, Logna F, Ali S, Philip PA, Sarkar FH. Re-expression of miR-200 by novel approaches regulates the expression of PTEN and MT1-MMP in pancreatic cancer. Carcinogenesis 2012;33:1563–71.

[27] Hom C, Lu J, Tamanoi F. Silica nanoparticles as a delivery system for nucleic acid-based reagents. J Mater Chem 2009;19:6308–16.

[28] Tang F, Li L, Chen D. Mesoporous silica nanoparticles: synthesis, biocompatibility and drug delivery. Adv Mater 2012;24:1504–34.

[29] Park IY, Kim IY, Yoo MK, Choi YJ, Cho MH, Cho CS. Mannosylated polyethylenimine coupled mesoporous silica nanoparticles for receptor-mediated gene delivery. Int J Pharm 2008;359:280–7.

[30] Yanes RE, Tamanoi F. Development of mesoporous silica nanomaterials as a vehicle for anticancer drug delivery. Ther Deliv 2012;3:389–404.

[31] Miller SA, Hong ED, Wright D. Rapid and efficient enzyme encapsulation in a dendrimer silica nanocomposite. Macromol Biosci 2006;6:839–45.

[32] Radu DR, Lai CY, Jeftinija K, Rowe EW, Jeftinija S, Lin VS. A polyamidoamine dendrimer-capped mesoporous silica nanosphere-based gene transfection agent. J Am Chem Soc 2004;126:13216–7.

[33] Lu J, Li Z, Zink JI, Tamanoi F. In vivo tumor suppression efficacy of mesoporous silica nanoparticles-based drug-delivery system: enhanced efficacy by folate modification. Nanomedicine 2012;8:212–20.

[34] Lu J, Liong M, Li Z, Zink JI, Tamanoi F. Biocompatability, biodistribution, and drug-delivery efficiency of mesoporous silica nanoparticles for cancer therapy in animals. Small 2010;6:1794–805.

[35] Hom C, Lu J, Liong M, Luo H, Li Z, Zink JI, et al. Mesoporous silica nanoparticles facilitate delivery of siRNA to shutdown signaling pathways in mammalian cells. Small 2010;6:1185–90.

[36] Chen AM, Zhang M, Wei D, Stueber D, Taratula O, Minko T, et al. Co-delivery of doxorubicin and Bcl-2 siRNA by mesoporous silica nanoparticles enhances the efficacy of chemotherapy in multidrug-resistant cancer cells. Small 2009;5:2673–7.

[37] Meng H, Liong M, Xia T, Li Z, Ji Z, Zink JI, et al. Engineered design of mesoporous silica nanoparticles to deliver doxorubicin and P-glycoprotein siRNA to overcome drug resistance in a cancer cell line. ACS Nano 2010;4:4539–50.

[38] Li X, Xie QR, Zhang J, Xia W, Gu H. The packaging of siRNA within the mesoporous structure of silica nanoparticles. Biomaterials 2011;32:9546–56.

[39] Chen Y, Chu C, Zhou Y, Ru Y, Chen H, Chen F, et al. Reversible pore-structure evolution in hollow silica nanocapsules: large pores for siRNA delivery and nanoparticle collecting. Small 2011;7:2935–44.

[40] Na HK, Kim MH, Park K, Ryoo SR, Lee KE, Jeon H, et al. Efficient functional delivery of siRNA using mesoporous silica nanoparticles with ultralarge pores. Small 2012;8:1752–61.

[41] Taratula O, Garbuzenko OB, Chen AM, Minko T. Innovative strategy for treatment of lung cancer: targeted nanotechnology-based inhalation co-delivery of anticancer drugs and siRNA. J Drug Target 2011;19:900–14.

[42] Ashley CE, Carnes EC, Epler KE, Padilla DP, Phillips GK, Castillo RE, et al. Delivery of small interfering RNA by peptide-targeted mesoporous silica nanoparticle-supported lipid bilayers. ACS Nano 2012;6:2174–88.

[43] Jeffs LB, Palmer LR, Ambegia EG, Giesbrecht C, Ewanick S, MacLachlan I. A scalable, extrusion-free method for efficient liposomal encapsulation of plasmid DNA. Pharm Res 2005;22:362–72.

[44] Morrissey DV, Lockridge JA, Shaw L, Blanchard K, Jensen K, Breen W, et al. Potent and persistent in vivo anti-HBV activity of chemically modified siRNA's. Nat Biotechnol 2005;23:1001–7.

[45] Mounkes LC, Zhong W, Cipres-Palacin G, Heath TD, Debs RJ. Proteoglycans mediate cationic liposome-DNA complex-based gene delivery in vitro and in vivo. J Biol Chem 1998;273:26164–70.

[46] Cullis PR, Krujiff B. Lipid polymorphism and the functional roles of lipids in biological membranes. Biochim Biophys Acta 1979;559:399–420.

[47] Liu Y, Liggitt D, Zhong W, Tu G, Gaensler K, Debs R. Cationic liposome-mediated intravenous gene delivery. J Biol Chem 1995;270:24864–70.

[48] Torchilin VP. Recent advances with liposomes as pharmaceutical carriers. Nat Rev Drug Discov 2005;4:145–60.

[49] Briane D, Slimani H, Tagounits A, Naejus R, Haddad O, Coudert R, et al. Inhibition of VEGF expression in A431 and MDA-MB 231 tumor cells by cationic lipid-mediated siRNA delivery. J Drug Target 2012;20:347–54.

[50] Gomes-da-Silva LC, Santos AO, Bimbo LM, Moura V, Ramalho JS, Pedroso de Lima MC, et al. Toward a siRNA-containing nanoparticle targeted to breast cancer cells and the tumor microenvironment. Int J Pharm 2012;434:9–19.

[51] Mokhtarieh AA, Cheong S, Kim S, Chung BH, Lee MK. Asymmetric liposome particles with highly efficient encapsulation of siRNA and without nonspecific cell

penetration suitable for target-specific delivery. Biochim Biophys Acta 2012;1818: 1633–41.

[52] Mishra S, Webster P, Davis ME. PEGylation significantly affects cellular uptake and intracellular trafficking on non-viral gene delivery particles. Eur J Cell Biol 2004;83: 97–111.

[53] Sato A, Choi SW, Hirai M, Yamayoshi A, Moriyama R, Yamano T, et al. Polymer brush-stabilized polyplex for a siRNA carrier with long circulatory half-life. J Control Release 2007;122:209–16.

[54] Glodde M, Sirsi SR, Lutz GJ. Physiochemical properties of low and high molecular weight poly(ethylene glycol)-Grafted Poly(ethylene imine) copolymers and their complexes with oligonucleotides. Biomacromolecules 2006;7:347–56.

[55] Gonzalez H, Hwang SJ, Davis ME. New class of polymers for the delivery of macro-molecular therapeutics. Bioconjug Chem 1999;10:1068–74.

[56] Bartlett DW, Davis ME. Physicochemical and biological characterization of targeted, nucleic acid containing nanoparticles. Bioconjug Chem 2007;18:456–68.

[57] Alabi C, Vegas A, Anderson D. Attacking the genome: emerging siRNA nanocarriers from concept to clinic. Curr Opin Pharmacol 2012;12:427–33.

[58] Kukarni RP, Mishra S, Fraser SE, Davis ME. Single cell kinetics of intracellular, nonviral, nucleic acid delivery vehicle acidification and trafficking. Bioconjug Chem 2005;16:986–94.

[59] Barlett DW, Davis ME. Impact of tumor-specific targeting and dosing schedule on tumor growth inhibition after intravenous administration of siRNA-containing nanoparticles. Biotechnol Bioeng 2008;99:975–85.

[60] Klein S, Zolk O, Fromm MF, Schrodl S, Neuhuber W, Kryschi C. Functionalized silicon quantum dots tailored for targeted siRNA delivery. Biochem Biophys Res Commun 2009;387:164–8.

[61] Li J-M, Zhao M-X, Su H, Wang Y-Y, Tan C-P, Ji L-N. Multifunctional quantum-dot-based siRNA delivery for HPV18 E6 gene silence and intracelular imaging. Biomaterials 2011;32:7978–87.

[62] Li J-M, Wang Y-Y, Zhao M-X, Tan C-P, Li Y-Q, Le X-Y, et al. Multifunctional QD-based co-delivery of siRNA and doxorubicin to HeLa cells for reversal of multidrug resistance and real-time tracking. Biomaterials 2012;33:2780–90.

[63] Lin Q, Chen J, Jin H, Ng KK, Yang M, Cao W, et al. Efficient systemic delivery of siRNA by using high-density lipoprotein-mimicking peptide lipid nanoparticles. Nanomedicine (Lond) 2012; http://dx.doi.org/10.2217/nnm.12.73.

AUTHOR INDEX

Numbers in regular font are page numbers and indicate that an author's work is referred to although the name is not cited in the text. Numbers in italics refer to the page numbers on which the complete reference appears.

SUBJECT INDEX

Note: Page numbers followed by "f" indicate figures, and "t" indicate tables.

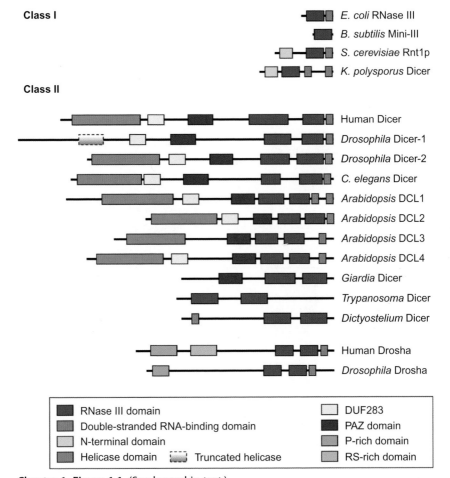

Class I

E. coli RNase III
B. subtilis Mini-III
S. cerevisiae Rnt1p
K. polysporus Dicer

Class II

Human Dicer
Drosophila Dicer-1
Drosophila Dicer-2
C. elegans Dicer
Arabidopsis DCL1
Arabidopsis DCL2
Arabidopsis DCL3
Arabidopsis DCL4
Giardia Dicer
Trypanosoma Dicer
Dictyostelium Dicer
Human Drosha
Drosophila Drosha

RNase III domain
Double-stranded RNA-binding domain
N-terminal domain
Helicase domain   Truncated helicase
DUF283
PAZ domain
P-rich domain
RS-rich domain

Chapter 1. Figure 1.1 (See legend in text.)

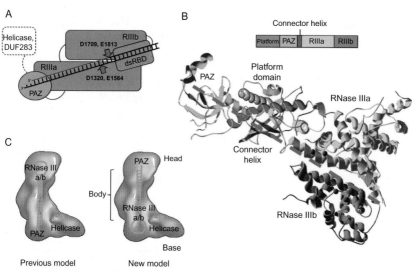

Chapter 1. Figure 1.2 (See legend in text.)

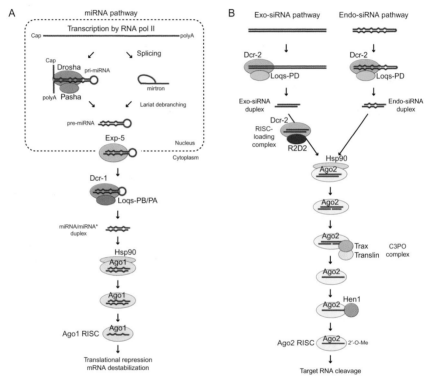

Chapter 2. Figure 2.1 (See legend in text.)

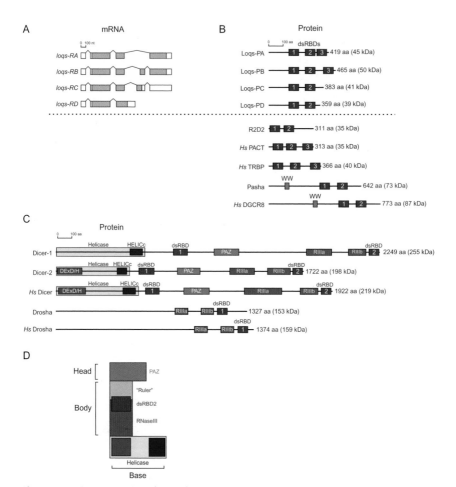

Chapter 2. Figure 2.2 (See legend in text.)

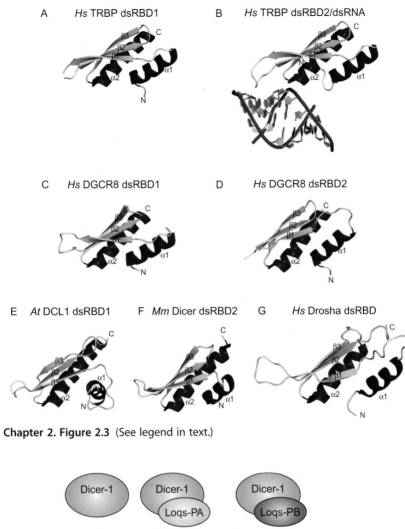

A   *Hs* TRBP dsRBD1

B   *Hs* TRBP dsRBD2/dsRNA

C   *Hs* DGCR8 dsRBD1

D   *Hs* DGCR8 dsRBD2

E   *At* DCL1 dsRBD1

F   *Mm* Dicer dsRBD2

G   *Hs* Drosha dsRBD

Chapter 2. Figure 2.3 (See legend in text.)

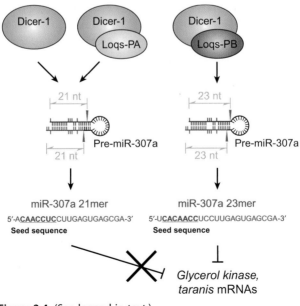

Dicer-1

Dicer-1 / Loqs-PA

Dicer-1 / Loqs-PB

21 nt

21 nt

Pre-miR-307a

23 nt

23 nt

Pre-miR-307a

miR-307a 21mer

5′-A**CAACCUC**CUUGAGUGAGCGA-3′

Seed sequence

miR-307a 23mer

5′-U**CACAACC**UCCUUGAGUGAGCGA-3′

Seed sequence

⊥ *Glycerol kinase, taranis* mRNAs

Chapter 2. Figure 2.4 (See legend in text.)

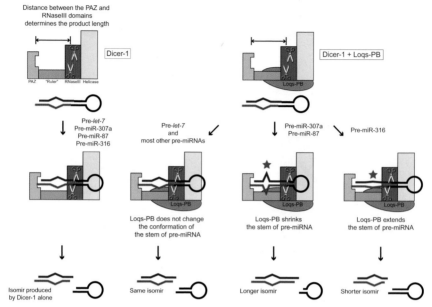

Chapter 2. Figure 2.5 (See legend in text.)

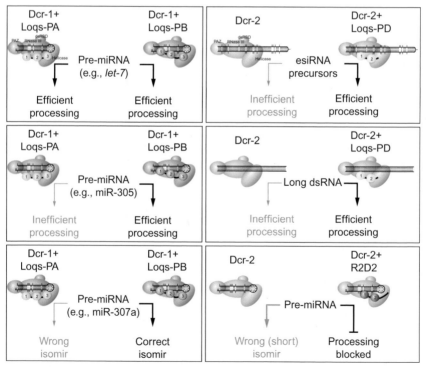

Chapter 2. Figure 2.6 (See legend in text.)

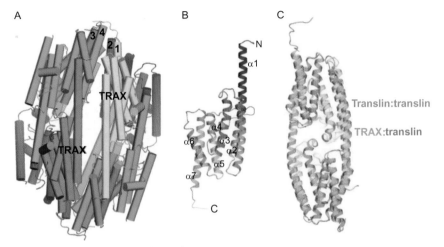

Chapter 3. Figure 3.3 (See legend in text.)

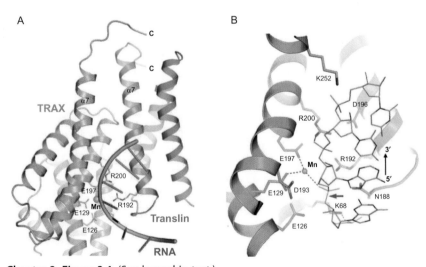

Chapter 3. Figure 3.4 (See legend in text.)

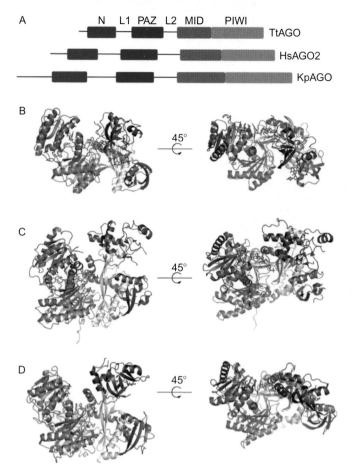

Chapter 4. Figure 4.1 (See legend in text.)

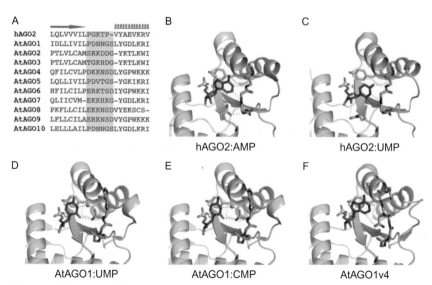

hAGO2	LQLVVVILPGKTP-VYAEVKRV
AtAGO1	IDLLIVILPDNNGSLYGDLKRI
AtAGO2	PTLVLCAMSRKDDG-YKTLKWI
AtAGO3	PTLVLCAMTGKHDG-YKTLKWI
AtAGO4	QFILCVLPDKKNSDLYGPWKKK
AtAGO5	LQLLIVILPDVTGS-YGKIKRI
AtAGO6	HFILCILPERKTSDIYGPWKKI
AtAGO7	QLIICVM-EKKHKG-YGDLKRI
AtAGO8	PKFLLCILEKKNSDVYEKSCS-
AtAGO9	LFLLCILAERKNSDVYGPWKKK
AtAGO10	LELLLAILPDNNGSLYGDLKRI

hAGO2:AMP

hAGO2:UMP

AtAGO1:UMP

AtAGO1:CMP

AtAGO1v4

Chapter 4. Figure 4.2 (See legend in text.)

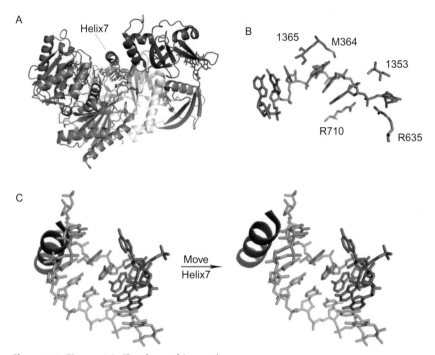

A Helix7

B 1365   M364

1353

R710   R635

C

Move
Helix7

Chapter 4. Figure 4.3 (See legend in text.)

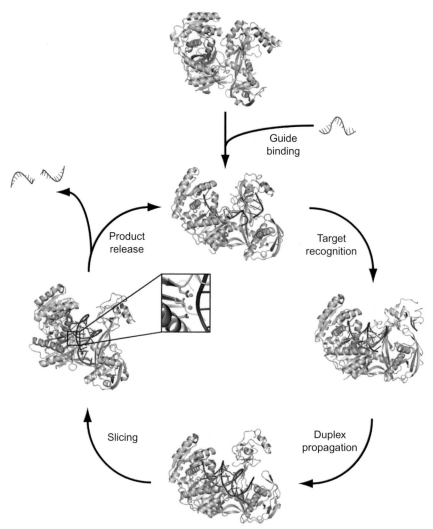

Chapter 4. Figure 4.4 (See legend in text.)

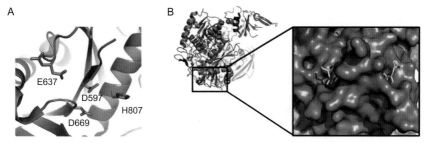

Chapter 4. Figure 4.5 (See legend in text.)

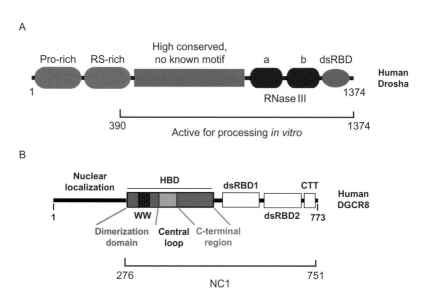

Chapter 5. Figure 5.1 (See legend in text.)

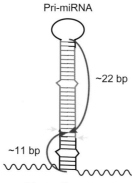

Chapter 5. Figure 5.3 (See legend in text.)

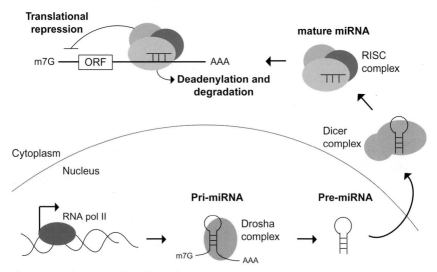

Chapter 6. Figure 6.1 (See legend in text.)

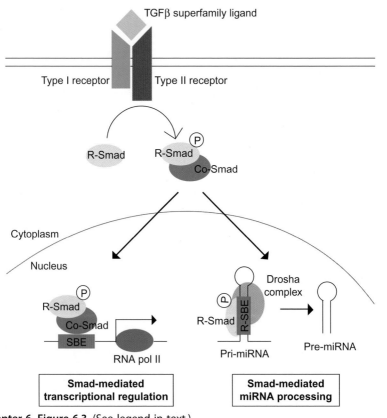

Chapter 6. Figure 6.3 (See legend in text.)

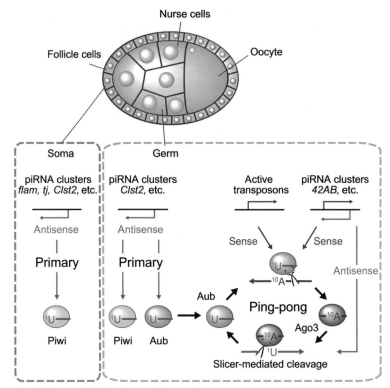

Chapter 7. Figure 7.3 (See legend in text.)

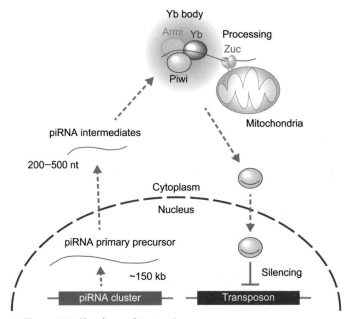

Chapter 7. Figure 7.5 (See legend in text.)

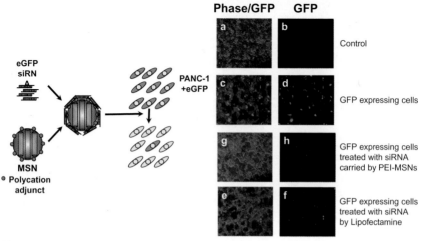

Phase/GFP   GFP

a	b	Control
c	d	GFP expressing cells
g	h	GFP expressing cells treated with siRNA carried by PEI-MSNs
e	f	GFP expressing cells treated with siRNA by Lipofectamine

Chapter 9. Figure 9.3 (See legend in text.)

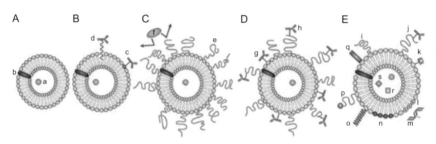

Chapter 9. Figure 9.4 (See legend in text.)

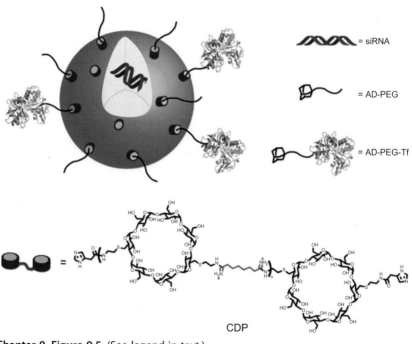

Chapter 9. Figure 9.5 (See legend in text.)